Developing Maths Lesson Planning and Frameworks

Developing Maths Lesson Planning and Frameworks

Mastery, Logic and Reasoning in the Classroom

*Yuqian Wang, Chris Brown,
Jeremy Dawson*

 Open University Press

Open University Press
McGraw Hill
Unit 4
Foundation Park
Roxborough Way
Maidenhead
SL6 3UD

Email: emea_uk_ireland@mheducation.com
World wide web: www.mheducation.co.uk

Executive Editor: Eleanor Christie
Editorial Assistant: Phoebe Hills
Content Product Manager: Graham Jones

British Library Cataloguing in Publication Data
A catalogue record of this book is available from the British Library

ISBN-13: 978-0-3352-5180-3
ISBN-10: 0335251803
eISBN: 978-0-3352-5181-0

Typeset by Transforma Pvt. Ltd., Chennai, India

Praise Page

"*Developing Maths Lesson Planning provides a clear and well-structured presentation of the potential of using the Causal Connectivity Framework in lesson planning to develop students' conceptual understanding and reasoning skills. Of particular significance is the weight given to comparative analysis of curricula and pedagogical approaches internationally. Whilst secondary maths teachers will find a wealth of approaches from which to draw, the text has broader significance for educators and school leaders in reflecting a wide range of topics in the contemporary discourse on pedagogy, assessment, mastery learning, practitioner development and change management.*"

Karen Taylor, Director of Education at Ecole Internationale de Genève, Switzerland and Associate Professor in Practice, Durham University School of Education, UK

"*In a time when mathematics education is more complex and challenging than at any other time, this book serves as a valuable guide for nurturing reasoning abilities. Through a blend of robust research and practical insights, the authors provide a deep and fresh perspective on effective mathematical reasoning. They explore the pedagogical aspects of mathematical reasoning as an integral component of teaching for mastery while offering guidance on its integration into lesson planning. The book also tackles the often overlooked yet vital aspect of assessing reasoning skills and offers suggestions for school leaders to support reasoning. I wholeheartedly recommend this book to teachers, school leaders and teacher educators.*"

Dr Holly Heshmati, Associate Professor in Mathematics Education at the Centre for Teacher Education, University of Warwick, UK

"*This book captures the essence of a new approach to lesson design and delivery that will inspire even the most accomplished practitioners to think deeply about what and how they teach mathematics. Peppered with real, practical examples that have already influenced teaching in a range of settings, it brings the Causal Connectivity Framework to life. The authors are powerful advocates of collaborative professional development as a means to mobilising sustained improvement. They argue, for attitudes and behaviours to advance at an individual classroom level teachers need support and encouragement to work with others to refine, augment and incorporate innovative strategies.*"

Wendy Truscott, former Senior Professional Development Adviser for Education Durham, Durham County Council, UK

"Developing Maths Lesson Planning and Frameworks: Mastery, Logic and Reasoning in the Classroom, provides much food for thought and includes many immediate 'take aways' to reflect on or try out. This book will help enhance any maths teacher's lesson planning, from the trainee teacher to the experienced practitioner; from those starting from scratch to plan a lesson to those utilising a ready-prepared lesson plan, perhaps written by someone else. I love that reasoning is explained as both a vehicle to understanding and understanding itself, and not simply a bolt on or plenary to a maths lesson. Reasoning is presented as being at the heart of successful mathematics teaching and learning; exactly where it belongs!"

Rose-Marie Rochester, Archimedes NE Maths Hub Lead/
BHCET Director of Maths, UK

"Developing Maths Lesson Planning and Frameworks considers the alignment of maths activities with a variety of reasoning styles, through an in-depth discussion of mathematical reasoning and proof. Drawing from research across the Globe, it considers the importance of lesson planning as a process of refining and reshaping pedagogy, through a mastery approach, by way of embedding reasoning into lesson planning. Through a comprehensive consideration of the Casual Connective Framework, it supports practitioners to integrate reasoning into their choices of classroom activities, and provides a range of examples of practical applications. A must have for current and future educators in Mathematics."

Pamela Burnip, Senior Lecturer of Mathematics ITT,
University of Sunderland, UK

"This book highlights the essential relationship between lesson planning and key mathematical elements such as mastery and reasoning. In particular, chapter 3: Lesson planning and the mastery approach, I believe will be an invaluable read for trainee teachers and those in the early stages of their career. The exemplification of ideas through useful classroom strategies, such as for trigonometry, gives the book a practical basis, as well as theoretical. I have no doubt that even experienced teachers will find this book enlightening and encourage them to re-evaluate elements of their practice."

Hayley Hands, Secondary PGCE Mathematics Lead,
Newcastle University, UK

Contents

Preface

This book is an exploration of teaching maths at the secondary stage. It summarizes an approach to lesson planning that recognizes the power of reasoning in how the journey of maths teaching and learning unfolds. It offers a simple thesis: lesson planning is not merely about anticipating how lessons are to appear on the surface, detailing, for example, the process by which teachers model approaches to topics for their students ('I-do', 'we-do', 'you-do'). Instead, lesson planning should be about presenting to students the logical foundations upon which activities are linked together. It is something tailored to the particular group of students and local circumstances. We propose the Causal Connectivity Framework (CCF) as the key means by which the logical process is made explicit. The CCF is a lesson planning regime aimed at developing conceptual understanding, which offers a way of organizing activities so that reasoning becomes inseparable from teaching and learning.

Background to this book

Behind this book lies a history of collaboration between the three authors. This history has involved extensive debating, questioning, and acknowledging of our own assumptions on teaching and learning in maths and more generally. As maths teachers, maths education researchers, and teacher educators in initial teacher training , a fundamental part of our work has been to understand what makes maths important to a person's life, to their successes, and even to their levels of contentment. Is it about problem-solving skills, being numerate, or recognizing patterns? Or is it perhaps about the reasoning lying behind these processes? We all addressed these issues in our independent work initially, before coming together to face them in collaboration.

Yuqian (Linda) Wang came to Durham University in 2012 (after becoming an award-winning practitioner during her 12 years of teaching maths in Shanghai) to pursue PhD research investigating the similarities and differences in understanding linear graphs between students in England and Shanghai (China). After that, she went into teaching in the north-east of England, where she met Wendy Truscott, the maths adviser for the Durham Local Authority. Wendy introduced Linda to Jeremy Dawson, the north-east coordinator for the Advanced Maths Support Programme, and this three-way collaboration led to various discoveries on how to teach some of the most 'tricky' maths topics at key stage 4, including turning points, trigonometry (see chapter 6), linear graphs, and vectors. Cohering around a postdoctoral project during the 2016–17 academic year, the work also involved the university maths education researcher Professor Adrian Simpson, the local Archimedes Maths Hub, and the heads of maths of six local schools. The project is titled, PANDA (Pedagogical Application of New Developments and Approaches, hereafter the PANDA project) pointed to our emphasis on finding *different* approaches to lesson design and reflecting on their delivery in local contexts. Successes here encouraged us to expand our exploration towards the primary level.

By the end of 2018, with Wang now an assistant professor at Durham University School of Education, she and Dawson had come to realize the importance of two particular elements for ensuring the sustainability of these investigations: (1) the need to develop an interpretative framework to explain in an explicit way how we design sequences of lessons, and why we do it in such a way; and (2) continuing to work with different educational institutions, at university and school level and with support of the local educational network. The first was a matter of reflection between Dawson and Wang, the results of which were the outline of the CCF we published in the Teaching Mathematics Journal in 2019. The second, however, required expert help in navigating the complexities of this network and in understanding how this network could be harnessed to help teachers further their thinking on their own. In 2019, Chris Brown joined Durham University as Research Director in the School of Education. He brought, in particular, expertise in harnessing professional learning networks to promote collaborative learning among teachers. The aim of this collaborative learning is to improve both teaching practice and student outcomes, not only in individual schools, but also in the school system more widely.

As a team of three, our approach has been built on consistent questioning of how to introduce the idea of reasoning to the classroom with cross-curriculum links, and how to nurture students' reasoning skills. We are particularly concerned with what teachers think of these efforts, and how they can be supported to develop further in this area. These starting points led to the creation of a small-scale pilot study funded by Durham University, on the topic of future-oriented continuing professional development, to promote reasoning in lower-primary settings. Around the same time, ongoing recovery from the COVID-19 pandemic period in 2021 carved out time for us to reflect again on the CCF and the most important aspects of maths lesson planning. The result of these reflections is the book we present to you now.

Reasoning with the Causal Connectivity Framework

Currently, teaching on reasoning is mainly concerned with reasoning within specific topic-related areas – for example, ratio and proportional reasoning, spatial and geometrical reasoning, reasoning about data, and reasoning about uncertainty. These reasoning areas are tied to particular maths content rather than being approached as skills to be developed as part of everyday maths learning. It is important to emphasize that there are different styles of reasoning, including generalizing, conjecturing, identifying a pattern, proving, convincing and arguing. We see this book as an opportunity to discuss ways of organizing maths activities and aligning them with this variety of reasoning styles.

During the lesson preparation stage, we argue that teachers should have a clear vision of the reasoning development running through their teaching, purposefully anticipating the shared and implicit reasoning established during the lesson – reasoning that students did not have access to at the start. The CCF we propose aims to develop reasoning in maths lessons through structuring activities in a particular sequence. Although most of the substantial examples and illustrations will be drawn from the projects that we have been involved in over the past seven years, we do not intend the consideration of the development of reasoning to start and end here. Instead, we present these excursions as inspiration to readers to reflect on their own conceptions of reasoning in the classroom.

The structure of the book

This book has three parts: Part 1 tackles mathematical reasoning in theory, Part 2 is on lesson planning and Part 3 looks at links between theory and practice, all preceded by an introductory chapter considering why reasoning is important, and why now. All of the contextual factors laid out in the introduction suggest that reasoning is heavily implicated in the future lives of students; indeed, reasoning is already included in the three aims in the national curriculum across all key stages in England.

There are further important dimensions to consider:

1 How the components of reasoning, or a framework for approaching it, are currently recognized internationally. This topic is addressed in Part 1.
2 The pedagogical aspects of reasoning, including its roles in predominant teaching approaches, such as teaching for mastery, and how reasoning can become an integral part of lesson planning, through the proposed CCF. This is the main theme of Part 2.
3 The assessment aspect of reasoning, and what school leaders can do to support reasoning, addressed in Part 3.

Within Part 1 (Mathematical Reasoning), chapter 1 discusses mathematical reasoning itself, and chapter 2 focuses on a particular form of reasoning, proof, drawing on examples of proof of circle theorems. In Part 2 (Maths Lesson Planning Frameworks), chapter 3 reflects on maths lesson planning as linked to mastery approaches; chapter 4 introduces the CCF in detail; chapter 5 looks at reasoning in a cross-curricular perspective, using probability from the transition between primary and secondary stages as an example; and chapter 6 draws from the trigonometry featured in the PANDA project. In Part 3 (In Between Theory and Practice), chapter 7 discusses the normal practice of assessing reasoning; and chapter 8 is about what school leaders can do to support reasoning and why it should be considered an urgent priority to do so.

Wang oversaw the book's completion. She wrote the introduction and chapters 1 to 7, and co-authored chapter 8 with Brown. Dawson contributed to chapters 2, 4 and 5.

Yuqian (Linda) Wang
December 2022 at Durham, England

Acknowledgements

This book would not have come to fruition without resilience, persistence and willingness in the sharing of knowledge and experience among us. We offer huge thanks to family, friends and colleagues for their support during the writing of this book. We particularly thank Dr Samuel Horlor, a long-standing friend, for being the first reader of the draft manuscript and for offering his insights. We thank our editor, Eleanor Christie, and her team for making this production happen.

Introduction

The next time you are walking into the maths classroom, take a few moments to think about your lesson planning, and perhaps quickly retrieve an example of having intentionally taught or used mathematical reasoning in this lesson. Before too long, some of students will probably appear in your mind, particularly those who might be among the most able or those who might need extra attention. So, we first step back and consider the lesson planning and mathematical reasoning from the perspective of the contemporary social environment – one in which learners grow up with digital learning media and in which they occupy digital worlds, i.e. 'where' learners are. Discussion of the values and ethos of this environment is followed by consideration of current trends in teacher education – 'what' they mean for the education landscape and for in-service teachers and trainees in particular.

With this broad-scale awareness, we look at global level, about recent changes in the focus of large-scale cross-national assessments such as the Programme for International Student Assessment (PISA) in maths. Then, these questions about 'what' we are dealing with turn into questions of 'how' to work in practice with the conditions we find. Our intention is to carefully strike a balance between remaining close to the comfort zone of teachers while at the same time trialling something new, in the form of the CCF for lesson planning. This book is an invitation for all in-service maths teachers and trainees to look beyond and behind your lesson planning, to exercise comparative judgement in considering how to structure lessons with mastery, logic and reasoning, and to make a realistic assessment of how to adapt the CCF in your local context, making it specific to the needs in your particular classroom and to the mathematics topics you teach.

Growing up in the digital world

In Prime Minister Rishi Sunak's first speech of 2023, he expressed the government's intention to ensure that all school pupils in England study maths in some form until the age of 18 (Department for Education (DfE), 2023). With this, the idea is for pupils to be better equipped for future jobs. Britain is currently at the beginning of the Fourth Industrial Revolution (Department for Digital, Culture, Media & Sport, 2017), a revolution involving a new digitally networked VUCA (volatility, uncertainty, complexity, and ambiguity) environment. Although reskilling and upskilling today's workers is essential under these circumstances, the need for reform in basic education is critical, if the next generations are to enter the workforce prepared for coming challenges.

More than 60 per cent of students in primary school today will end up working in job types that do not yet exist (OECD, 2019a). This should not be taken as a reason to panic, however. Partly because students currently in primary school have grown up in the digital world, with tablet interactivity, video chat, etc. as part of their lives from very early childhood (Barr, 2019). Equally importantly, there is more and more awareness of future challenges in the education community, exemplified not least in efforts to promote and develop digital literacy. Scotland, for example, launched its Digital Learning and Teaching Strategy in 2016 (Education Scotland, n.d.), its aim to maximize opportunities for all students to engage with digital technology. The strategy emphasizes the educator's skills and confidence, making digital technologies a central consideration of the curriculum and assessment at the delivery stage, and empowering leaders to facilitate and lead innovation. Digital learning and teaching are clearly more than a matter of students accessing digital devices.

When the COVID-19 pandemic hit in 2020 and the nation went into lockdown, scholars in Scotland (just as everywhere else) had ready opportunity to reflect on the digital technology in the education landscape. Lockdown forced changes in practice as school buildings were closed, and assumptions about the teaching and learning process, and about teachers' roles, have been challenged (Brown et al., 2021). Technology's position in these processes has risen from being merely a third space to being *the* space. Third space is a metaphorical way of referring to ways of learning, beyond traditional institutional context, with their curricula and rigorous schedules – one example being mobile learning (Schuck, Kearney and Burden 2016). The pandemic period challenged the drawing of fundamental distinctions between schools (the institution as the first space), learning experience (social networks as the second space), and technological places (as third space, an interaction between the first two). The question is what this development means to us?

Key to teachers tackling associated challenges is transforming their understanding of 'knowing' and, linked to this, recognizing shifts in boundaries brought about by the new ways teachers and students interact (Colville et al., 2021). This means acknowledging that both teachers and students are active learning participants. Typically, teachers are thought of as carriers of 'knowledge' – mediators between students and the syllabus learning requirements for the subject. But imagine if the teacher's role were rethought so that they become facilitators of reflection. This would involve encouraging teachers to orient themselves to reflection and to make reflection explicit in the teaching and learning process. To do so, lesson planning is vital. One of the main contributions of the CCF (see chapter 4) is to establish grounds for consensus among trainee and in-service teachers about what lesson planning in mathematics should look like.

Better mathematics skills

One assumption often made is that to live in this ever-changing society we need different skills, those which the Organization for Economic Cooperation and Development (OECD) calls 'future skills'. We now narrow things down to the mathematics education field: how is this need integrated, what are future mathematics skills, and what lies behind these skills?

One possible focus for the future of mathematic learning is problem-solving, especially when it is related to non-routine problems. Problem-solving strategies used by students and trainee teachers have included those of trial-and-error, making a drawing, and guess-check-revise (Barham, 2020). A potential argument is that the more strategies mastered, the more chance there is for success. So, does it simply mean the more strategies the merrier?

Exploration has also focused on flexibility in changing strategies if one does not work. Let's look at flexibility in practice, considering, for example, how well primary students respond to different problems – their inter-task flexibility, and how they respond within one problem – their intra-task flexibility (Elia, Heuvel-Panhuizen and Kolovou, 2009). One of the findings is how strategies chosen by students may be related to the way the topic is introduced in the curriculum. In the case of function, there are two ways of introducing the topic at secondary level: (1) a mapping view, which emphasizes domain, range, and correspondence, a symbolic manipulation of expression; and (2) a correspondence approach, which starts with the function relations between quantities, such as function machines, or with a graphic qualitative description (Wang, 2015). Each approach connects the explicit expressions and process of change, which are crucial to the nature of function, in different ways, but each offers further complexities (Yerushalmy, 2000). In other words, there is no one best approach. So as teachers, we should always be aware of how the topic unfolds in our curriculum. Chapters 5 and 6 will give two examples to explore this.

Let's go back to problem-solving. To solve a problem requires the problem to be analysed scientifically. But does scientific analysis imply promotion of critical thinking? Another direction in which educational expectations have moved in the twenty-first century is exemplified in the 4Cs in the Indonesian curriculum: creative thinking, critical thinking, communication, and collaboration. With the curriculum pointing to critical thinking, how can students' learning best be organized to improve this aspect?

Pahrudin and his colleagues (Pahrudin et al., 2021) compared two models, the STEM-inquiry model and the STAD (student team achievement division) model, to investigate which one promoted critical thinking better, and to ascertain possible reasons. The STEM-inquiry model follows a systematic order of steps: problem orientation and questions, creating a hypothesis, planning and conducting an investigation, analysing and interpreting, concluding and communication. The STAD model, on the other hand, starts from goals and motivations behind them, highlighting group division, presentation by the teacher, learning as a team, evaluation through quizzes, and appreciation for the group's achievement. The research found that the STEM-inquiry model has more positive impact on critical thinking, and this is because every step is based on students' previous learning experience, with a deeper connection added at each step. Chapter 4's discussion of the CCF provides detail on how these deeper connections can be pursued.

Metacognitive skills – 'thinking about thinking' – have also featured in the OECD's vision for future skills (OECD, 2019b, p. 89). This refers to a process of understanding one's own thinking, a process involving the elements of reasoning and reflection. The Smith Report (DfE, 2017), a review of 16–18 mathematics education in the UK, emphasizes the centrality of developing analytical reasoning for success in the world of the future. These calls go far beyond the requirements of numeracy skills. In Part 1, we will look at mathematical reasoning in detail as part of this move to view lesson planning as a process of reasoning.

References

Barham, A.I. (2020) Investigating the development of pre-service teachers' problem-solving strategies via problem-solving mathematics classes, *European Journal of Educational Research*, 9(1): 129–41. Available at: https://files.eric.ed.gov/fulltext/EJ1241227.pdf (accessed 6 September 2023)

Barr, R. (2019) Growing up in the digital age: early learning and family media ecology, *Current directions in psychological science*, 28(4): 341–6. https://doi.org/10.1177/0963721419833824

Brown, J., McLennan, C., Mercieca, D. et al. (2021) Technology as thirdspace: teachers in Scottish schools engaging with and being challenged by digital technology in first COVID-19 lockdown, *Education Sciences*, 11(3): 1–16, 136. https://doi.org/10.3390/educsci11030136

Colville, T., Hulme, S., Kerr, C. et al. (2021) Teaching and learning in COVID-19 lockdown in Scotland: teachers' engaged pedagogy, *Frontiers in Psychology*, Vol. 12. https://doi.org/10.3389/fpsyg.2021.733633

Department for Digital, Culture, Media & Sport (2017) *The Fourth Industrial Revolution*. Available at: https://www.gov.uk/government/speeches/the-4th-industrial-revolution (accessed 21 June 2023).

Department for Education (2017) *Smith review of post-16 mathematics: report and letter*. Available at: https://www.gov.uk/government/publications/smith-review-of-post-16-maths-report-and-government-response (accessed 21 June 2023).

Department for Education (2023) *Studying maths to 18 – what you need to know*. Available at: https://educationhub.blog.gov.uk/2023/01/04/studying-maths-to-18-what-you-need-to-know/ (accessed 3 June 2023).

Education Scotland (n.d.) *Digital Learning and Teaching Strategy for Scotland*. Available at: https://www.gov.scot/policies/schools/digital-learning-and-teaching/ (accessed 07 June 2023).

Elia, I., van den Heuvel-Panhuizen, M., and Kolovou, A. (2009) Exploring strategy use and strategy flexibility in non-routine problem solving by primary school high achievers in mathematics, *ZDM Mathematics Education*, 41(5): 605–18. https://doi.org/10.1007/s11858-009-0184-6

OECD (2019a) *TALIS 2018 Results (Volume I): Teachers and school leaders as lifelong learners*, TALIS. Paris: OECD Publishing. https://doi.org/10.1787/1d0bc92a-en

OECD (2019b) *OECD future of education and skills 2030*. Available at: https://www.oecd.org/education/2030-project/teaching-and-learning/learning/learning-compass-2030/OECD_Learning_Compass_2030_Concept_Note_Series.pdf (accessed 12 July 2023).

Pahrudin, A., Alisia, G., Saregar, A. et al. (2021) The effectiveness of science, technology, engineering, and mathematics inquiry learning for 15-16 years old students based on

K-13 indonesian curriculum: the impact on the critical thinking skills, *European Journal of Educational Research*, 10(2): 681–92. Available at: https://files.eric.ed.gov/fulltext/EJ1294656.pdf (accessed 6 September 2023).

Schuck, S., Kearney, M., and Burden, K. (2017) Exploring mobile learning in the third space, *Technology. Pedagogy and Education*, 26(2): 121–37. https://doi.org/10.1080/1475939X.2016.1230555

Wang, Y. (2015) *Understanding linear function in secondary school students: a comparative study between England and Shanghai.* [Doctoral thesis, Durham University]. ProQuest Dissertations and Theses Global. Available at: http://etheses.dur.ac.uk/11230/ (accessed 18 May 2023).

Yerushalmy, M. (2000) Problem solving strategies and mathematical resources: a longitudinal view on problem solving in a function based approach to algebra, *Educational Studies in Mathematics*, 43(2): 125–47. https://doi.org/10.1023/A:1017566031373

Part 1

Mathematical Reasoning

1 Mathematical reasoning itself

Key arguments

In this chapter, we argue that reasoning should serve as a basis for mathematics teaching and learning, as it unifies understanding, problem-solving, and fluency. The chapter establishes a set of rules – for example, generalization by explanation, proof by conjecturing, justification by mathematizing – for how to think about mathematics concepts in a way that is in continual flux.

Key terms

The chapter summarizes three terms related to types of reasoning that can be embedded into lesson planning. We see these three processes as the building blocks of reasoning.

- Generalization by explanation
  A key part of reasoning is generalization. We introduce one important way to generalize: using a series of activities to enable students to discover mathematical structures, starting in the empirical realm and then moving to the abstract level. This constitutes a mode of explanation aimed at understanding mathematical objects and validating claims made.

- Proof by conjecturing
  The next key step in the reasoning process is proof. Creating written justifications is about organizing mathematics objects in a rigorous and formal way with the aim of developing reasoned conjectures. This type of reasoning builds upon the processes of structural generalization.

- Justification by mathematizing
  Finally, we introduce the processes of justification, which develop from intuitive to structured kinds. These processes generate arguments mathematically.

This chapter starts by outlining the global context of how mathematical reasoning is conceptualized, with reference to two large-scale international student assessments: PISA and Trends in International Mathematics and Science Study (TIMSS). Then it focuses on three aspects of reasoning: generalization by explanation, proof by conjecturing, and justification by mathematizing. Finally, it discusses two kinds of reasoning: that associated with particular mathematical topics, such as spatial geometric reasoning, and that not associated with particular topics, such as creative reasoning.

1.1 Starting from mathematics and mathematics reasoning

The nature of mathematics is initially debated in Greek philosophical thought. Contrasting views are put forward by Plato and his student Aristotle. Plato held a rationalist or absolutist view of mathematics, seeing it as pure mathematics, an abstract mental activity that is separate from the senses. The nature of mathematics pursuits is to seek external, independent and existing truth. This absolutist philosophy of mathematics searches for rigid and logical structures in mathematical systems, independent from culture and values (Ernest 1991).

On the other hand, Aristotle took the experimentalist (or fallibilist) view, arguing for the importance of experiments, observations and discoveries in finding new truths (Wang 2015). This view emphasizes processes, with mathematics linked to social practices in recontextualizing and reproducing knowledge. Ernest (1997, p. 3) argues that under this view, 'mathematics is experienced as warm, human, personal, intuitive, active, collaborative, creative, investigational, cultural, historical, living, related to human situations, enjoyable, full of joy, wonder, and beauty'. This view has often been propagated to encourage those pursuing mathematics to feel a sense of belonging. The views of Plato and Aristotle reflect two important sides of mathematics in schools: mathematics itself in the curriculum, and the process of learning mathematics.

Within this wider picture of mathematics learning, the purpose of mathematical reasoning is depicted as being to allow students to 'arrive at results that they can fully trust to be true in a wide variety of real-life contexts. It is also important that these conclusions are impartial, without any need for validation by an external authority' (OECD n.d.).

In the past, educators and students faced fixed expectations from the national curriculum and the standardization of assessments in their daily classrooms. Mathematical reasoning was set as an objective to achieve, one with a collection of categorizations as sub-objectives. Today, we face a quite different set of understandings, comprising an attitude to mathematical reasoning, a process of mathematical reasoning and a secure pedagogy for fostering mathematical reasoning. In this book, we adopt the fundamental position of Aristotle, which holds that mathematical reasoning is not merely a logical procedure but what happens based on this logical procedure (an idea we will discuss in the next chapter). It is a conversation, a process of argument. This process of argument (either in oral or written form) is as important as the solution of the problem; the answer as an outcome, or the product of the truth-seeking.

Reasoning is considered central among the key skills that the next generation of citizens will need if they are to thrive in their future lives and careers. It has been particularly emphasized in national curricula in mathematics around the world – for example in Australia (McCluskey, Mulligan and Mitchelmore 2016) and England (DfE 2021). The two large-scale cross-national assessments, PISA and TIMSS, have both identified reasoning as a vital element. The way mathematical reasoning is perceived in these two assessments presents two different approaches to maximizing reasoning in teaching and learning. The PISA 2022 mathematics framework emphasizes both deductive and inductive reasoning for 15-year-old students (OECD n.d.), in the form of reasoning within problem-solving (without being associated with particular topics); while TIMSS has focused on reasoning around particular areas in the curriculum when assessing Grade 4 and Grade 8 students (TIMSS 2019) – reasoning within mathematical topics.

1.1.1 Mathematical reasoning in PISA

The PISA 2022 framework describes mathematical reasoning as an 'ability to reason logically and present arguments in honest and convincing ways' (OECD n.d.). This notion stems from the very nature of mathematics in the OECD's view: 'Mathematics is a science about well-defined objects and notions that can be analysed and transformed in different ways using "mathematical reasoning" to obtain certain and timeless conclusions' (OECD n.d.). Its version of mathematical reasoning is divided into two types: 'mathematical (or deductive)' and 'statistical (or inductive)' (OECD 2018, p. 3). These two reasoning types are unified by three mathematical problem-solving processes:

1 Formulate: identify the mathematical variables in a problem and where to use mathematics; transform this into mathematical treatments, implying mathematical structures and representations.
2 Employ: use mathematical concepts and procedures to derive solutions.
3 Interpret and evaluate: reflect on the solutions and evaluate them in relation to the context of the problem.

In PISA, the mathematical reasoning and problem-solving processes are two separate elements, understood as reasoning mathematically and solving contextual mathematical problems, respectively. So, the validity of mathematical reasoning lies in the domain of real-life problems. In other words, every instance of problem-solving is a hidden exercise in mathematical reasoning, and an attempt to establish a mathematical approach. In the next section, we examine the TIMSS framework for reasoning since this employs the contrasting focus of curriculum-based assessment.

1.1.2 Mathematical reasoning in TIMSS

TIMSS (2019) lists reasoning as one of the three cognitive domains in mathematics (the other two being knowing and applying). At its Grade 8 assessment, reasoning falls within the so-called 'inner mathematics context' – mathematics without a real-life context (Wu 2009), and it implies confronting 'unfamiliar situations, complex contexts, and multistep problems' (TIMSS 2019, p. 22). TIMSS refers to reasoning mathematically as 'logical systematic thinking', something which 'includes intuitive and inductive reasoning based on patterns and regularities' (TIMSS 2019, p. 24). However, TIMSS also requires that certain major contents be tested, i.e., number, algebra, geometry, data and probability at Grade 8. Within this basic knowledge and its related skills in mind, problem-solving is what provides the essential direction for mathematical argument, through students (1) knowing which procedures to carry out, (2) applying what they already know to create representations for communicating their solutions, and (3) reasoning, such as through making conjectures and logical deductions.

In this way, the problem-solving process is no longer a process of solution-seeking imposed on students by assessment, something in which they have to demonstrate their academic abilities in mathematics. It becomes instead a matter of students fostering logic based on their assumptions, reflecting on these assumptions, and justifying themselves – all taking place within the context of these mathematical areas. Hence, reasoning itself in the TIMSS mathematics framework has been described as a matter of six steps (TIMSS 2019, p. 24):

- 'Analyse': determine, describe, or use relationships among numbers, expressions, quantities, and shapes.
- 'Integrate/synthesize': link different elements of knowledge, related representations, and procedures to solve problems.
- 'Evaluate': evaluate alternative problem-solving strategies and solutions.
- 'Draw conclusions': make valid inferences on the basis of information and evidence.
- 'Generalize': make statements that represent relationships in more general and more widely applicable terms.
- 'Justify': provide mathematical arguments to support a strategy or solution.

1.1.3 Summary

In summary, there is an important lesson to learn from these two international assessments. The power of reasoning comes from recognizing its relevance to the mathematical learning process and to the problem-solving process. It is not a matter of right or wrong, but of collectively moving forward in our relationship with mathematics itself, and with mathematics in its real-world applications. The key to success in mathematics is a full and ongoing integration of different aspects of reasoning in the teaching and learning process. By welcoming and honouring reasoning in schools, in both problem-solving contexts and the mathematics domain itself, teachers and researchers play the most important role as we incorporate it into the directions we set for mathematics learning. Putting all this together, we arrive at the conclusion that reasoning is needed even more than conceptual understanding and fluency; in relation to reasoning, everything else in mathematics learning has only second-rate value.

Since we have already hinted that there is more than one way to engage in mathematical reasoning, in the next section, we focus more on *which* version of mathematical reasoning to take forward.

Box 1.1 Problem-solving and mathematical reasoning – which do you prefer?

1 Based on PISA: Every instance of problem-solving is a hidden exercise in mathematical reasoning, and an attempt to establish a mathematical approach.
2 Based on TIMSS: Problem-solving is a matter of students fostering logic based on their assumptions, reflecting on these assumptions, and justifying themselves.

1.2 Which mathematical reasoning?

The types of reasoning offered by PISA and TIMSS represent by no means the only ways of interpreting the notion, and many nuances can be found in academic literature on the subject. The examinations of PISA and TIMSS do, however, point to one particularly important duality shaping the discussion: domain-specific reasoning versus domain-general reasoning.

Reasoning is viewed as one of the two elements (with problem-solving) in mathematical thinking (Goos and Kaya 2019). The term mathematical reasoning is often closely related to problem-solving, especially for problems that are not used routinely in the classroom (Nunes et al. 2015), and to particular mathematical areas (Watson, Jones and Pratt 2013). The difference between them is not like the logical process involved in proving different theorems – such as Pythagoras' theorem or circle theorems – or congruency of triangles. Rather, there are two different ways of conceptualizing reasoning, as the differences between PISA and TIMSS illustrate. In empirical research, this has led to two particular kinds of reasoning commonly being addressed (Hjelte, Schindler and Nilsson 2020): domain-specific and domain-general reasoning. Domain-specific reasoning is something akin to a particular topic and method. Domain-general reasoning is more like an ideology. In this section, we explore these two approaches from the perspective of their academic interpretations.

1.2.1 Reasoning in a mathematical way

In practice, when we think about mathematics itself, we tend not only to explain 'what' and 'how' in an effort to strengthen and deepen students' conceptual understanding. But we also attempt to establish the basis of the reasoning on which all the concepts and procedures depend. We take two examples from observing Year 8 lessons taught by two of our teacher trainees to argue that reasoning is the foundation of mathematics learning for both conceptual and procedural understanding. The examples show how trainees consciously deploy domain-specific reasoning to promote understanding. At the same time, we see that the trainees themselves recognize the need for reasoning to be acknowledged as a separate issue, rather than simply as a step in supporting students' understanding.

Classroom Observation 1: teaching the standard form $a \times 10^n$ At lower-secondary level, where a is a number bigger than or equal to 1 and less than 10. The lesson starts from explaining the need for standard form; for example, writing a large number like a googol requires a 1 followed by one hundred zeros: 10^{100}. Then, the explanation goes on to tackle why 10 is the basis of standard form rather than, for example, 2 or 3. It is because 10 is at the heart of our whole number system, through links to the concept of place value. This explanation supports students' understanding through connections, drawing a logical link between standard form and place value.

Classroom Observation 2: solving a linear inequality With an unknown on one side, such as $x + 5 > 4$. The lesson starts from conceptual understanding of inequality symbols (greater than, less than, greater than or equal to, etc.) and then illustrates, using a number line, potential difficulties in understanding these symbols in relation to real numbers. The reasoning part emerges here: 'If g is less than 5, the largest number that g can be is 4. Do you agree? Explain your answer.' As some students might only consider integers rather decimals, it is natural to turn to the number line to explain. This supports their understanding of the number system, progressing from integers to rational numbers.

Then, the lesson moves on to the procedure itself, solving inequalities such as $x + 6 > 2$, progressing to $2p + 5 > -1$. It does so by recalling the procedure as outlined above. However, when meeting a negative coefficient, such as $-2x \geq 4$, the question is whether students should be asked to memorize the generalized rule that multiplying or dividing both sides by a negative number reverses the inequality, without understanding why this should be so. Alternatively, can we explain it? The trainee in our lesson observation tried to establish a reasoning process: $-2x \geq 4$, so by adding $2x$ on both sides, we get $0 \geq 4 + 2x$. Then by subtracting 4 on both sides, we get $-4 \geq 2x$, and then dividing by 2 on both sides arrives at the answer $-2 \geq x$, which can finally be rewritten as $x \leq -2$. To explain both the generalized rule and why the rule works – even for lower ability students – seems plausible from the point of view of deepening their procedural understanding.

Reflection These two examples sought not right-wrong answers but a structural perspective on understanding – not memorization but a functional approach. They recognized that mathematics teaching is not about mastering facts and practice, but about offering dynamic relationships between the mathematical contextual constraints and teaching opportunities. In theory, this is explained as moving through the zone of promoted action (ZPA) and the zone of free movement (ZFM) to reach the zone of proximal development (ZPD) (Blanton, Westbrook and Carter 2005). It is not simply a matter of whether students can do the questions by the end of the lesson, but rather of building in time for the formation and elaboration of knowledge. It is better, simply, to prioritize reasoning. As we continue to build up a suitably full picture of reasoning, next we examine three aspects of reasoning as it is tackled in research.

Generalization by explanation in research How reasoning develops is profoundly affected by abilities in discovering underlying mathematical structures. This is revealed clearly in the processes of generalization, since certain special cases apply to a large set of mathematical objectives. In the process of discovery, two focuses have been identified: mathematical objectives and the validation of claims made. Mathematical objectives are fact-based, while the process of the validation has two elements: the empirical – empirical generalization – and the theoretical – structural generalization. There is also a process for shifting from one to the other (Carraher, Martinez and Schliemann 2007). In the case of functional relations between variables, it has been suggested that the concept of function is best introduced 'as input-output mappings' rather than as a sequence (Carraher, Martinez and Schliemann 2007, p. 20). The shift from empirical to theoretical generalization is achieved by the realization of a finite set of ordered pairs, such as a table of data, to a general rule-based infinite domain, such as using 'n' to represent any number. This focus supports students in acting intentionally and purposefully towards mathematizing. But how?

The most important thing to understand about generalization is the shifting process. Zayyadi and Kurniati (2018) argue that a trial-and-error strategy does not help in finding generalized mathematical patterns – a functional relationship – but only to identify a reasonable pattern, i.e. a local recursive relationship. Perhaps the crucial question we need to ask is not whether to 'shift', but what motivates this shift. Through the investigation of the views of primary pre-service teachers, it appears that three elements lead to such a shift: (1) prior experience with similar activities, (2) alternative approaches and (3) peer collaboration (Alderton et al. 2018). Let's take a glance at the second element in particular: alternative approaches.

An example of generalization by explanation: reasoning about data A good place to begin is the case of reasoning about data. When teaching statistics in lower-secondary settings, teachers might rely on contexts that students are familiar with; for example, collecting data from real life, representing data in tables and diagrams such as pie charts, bar charts and cumulative frequency diagrams, and analysing data to reach statistical conclusions. Reasoning about data, which is mainly associated with statistics education, is fundamentally based on how to interpret given data, such as medians, quartiles and box plots (Bakker, Biehler and Konold 2004), as well as distribution, variation, comparing groups, etc.. (Biehler et al. 2018). Alongside the development of technology, graphical visualization can also act as intermediate representation

(Gutiérrez et al. 2008). This, however, does not mean we can skip or suppress the interpretation process. During this process, students might have their own interpretations, but teachers as well as students need to remain constantly vigilant about alternative interpretations underlying how the statistical concepts are understood and perceived (Doerr and English 2003), as well about statistical thinking (Ben-Zvi 2004).

With this in mind, we move beyond a debate about whether all students in any given class have necessary prior experience with data collection. We move instead towards considering the students' experience of the present. Wilkerson and Laina (2018) describe how to use local publicly available datasets to promote reasoning in Grade 5 and 7 classrooms: repurposing data via *storytelling*. Data modelling arises when transferring data from their original context to the context of the learning experience. Then, different interpretations of the same dataset can lead to different paths of inquiry – one example being storytelling – in such a way that emergent, more purposeful external representations are produced. These stories do not have right or wrong answers, but the way in which alternative approaches act upon each other creates a blended and shared experience for students.

Proof by conjecturing in research It is difficult to overstate the pervasiveness of proof in mathematical reasoning. Stylianides (2008) proposed that proof means a valid argument (a connected sequence of assertions that the mathematics field has agreed upon) based on accepted truths (theorems, definitions, etc.) for or against a mathematical claim. In a nutshell, there is a difference between structural generalization (i.e. identifying a definite pattern which is a possible mathematical way of solving the problem) and empirical generalizations (i.e. identifying a plausible pattern for which a counter-example can be found) (Stylianides 2008). To bridge between the empirical and the structural, students need to attune with the conjunctures, with what could be true in their hypothesis to make a justification. Herbst (2004) called the process *reasoned conjectures.*

An example of proof by conjecturing: spatial and geometrical reasoning In this context, reasoning is mainly related to geometry topics with diagrams, such as transformations from 2-D to 3-D and the interrelationships between them (Mulligan et al. 2020), mental rotation, spatial orientation and spatial visualization (Lowrie et al. 2021). In the process of making 'reasoned conjectures', student interaction with diagrams plays a particularly important role in producing the hypothesis that leads to judgements as to whether a potential conjecture is reasonable (Chen and Herbst, 2013). Two perspectives are at play: the diagram itself and interactions with it, meaning that there exists a state of unity between different systems. Chen and Herbst (2013) theorized kinds of interaction as: (1) reasoning with and through semiotic systems; for instance, an operational way to make sense of the diagram by labelling the components of the shape; (2) gesture as a meaning-making symbolic system; for instance, a perceptual way to recognize properties; and (3) the modal system of language as a resource for making statements. How these three kinds of interaction reveal their dynamic interconnectedness is at the heart of configuring and/or refiguring the diagram itself.

Justification by mathematizing in research Justification pertains to proof in a more generalized sense. The main words used in justification are 'demonstrate, justify, prove, show' and similar (Mizrahi 2020). This means that proof is one form of justification, but not all justifications are proof (Brodie 2009). Furthermore, justification means not only to lay the foundations for proof, but also to cultivate a more robust connection between conceptual understanding and logical reasoning (Eko, Prabawanto and Jupri 2018). Interestingly, Martino and Maher (1999) have discovered that leading students to consider similarities and differences between conceptual understanding and logical reasoning through reflecting on their prior experience brings new emphasis to the processes of justification, or partial justification. Justification is accumulated by explanation supported with deeper understanding of mathematical concepts – mathematizing. But when Brodie (2010) observed an algebra reasoning Grade 10 class in Johannesburg, South Africa (asking if $x^2 + 1$ can equal 0 if x is a real number), it appeared there was a gap between learners' intuitive justifications and the ability to express these justifications in a structured and meaningful way. This serves as a fundamental insight into the nature of generating arguments mathematically, in both oral and written forms of justification leading to formal proof (see chapter 2).

Levels of justification Common practice around justification in the classroom is to ask students to justify why a mathematical statement or solution is always, sometimes, or never true. Students can use patterns, structures, representations, or even list counter-examples to refute or agree. Lo, Grant and Flowers (2007, p. 12) have proposed that there are five levels within written justification:

1 Level 4: justification is clear and mathematically correct.
2 Level 3: justification is mostly clear and mathematically correct. Students may have glossed over or omitted some aspects of the justification.
3 Level 2: parts of the justification are mathematically incorrect or contain insufficient details.
4 Level 1: justifications are mainly descriptive or illustrative of the steps.
5 Level 0: written work is missing or does not contain a valid reasoning strategy.

There is a lot to unpack around these levels. First, they offer teachers clues to discern if students are struggling with written justification itself or understanding of the relevant concepts, or both. Secondly, rather than describing or illustrating mathematical elements through, for instance, words or pictures to show their understanding, students are driven to understand or coordinate these elements into different steps as they formulate a written justification.

Thirdly, students may also fall into the trap of inner-justification concerning how these steps are linked with each other – why one follows another.

Widjaja et al. (2020) suggested that students need to be equipped with strategies to refute, and then they need to use argument to confirm or refute. Four levels of justification were identified (Widjaja et al. 2020, p. 339):

1 Level 4: using a generalizable argument.
2 Level 3: using an argument to confirm or refute a conjecture.
3 Level 2: using counter-examples to refute a conjecture.
4 Level 1: using examples and searching for counter-examples to explain or verify a conjecture of generalization
5 Level 0: no justification (including appealing to the authority of others).

Box 1.2 Reasoning

1 Generalization by explanation: be aware of what motivated its shifting process, from empirical to structural generalisation – an alternative approach for your subject knowledge development.
2 Proof by conjecturing: express these justifications in a structured and meaningful way, e.g. a formal written proof (see chapter 2).
3 Justification by mathematizing: such as the activity of 'Always, sometimes, or never true'.

In summary, we have uncovered how mathematical reasoning develops its own forms of generalization by explanation, proof by conjecturing and justification by mathematizing in research. This recognizes the importance of reasoning in the mathematics domain as a method to follow or to promote. Now we launch our exploration into the question of what the perceived goals in the mathematics classroom are – on the presumption that commitment to mathematics reasoning is also a form of ideology. In other words, we are moving to domain-general reasoning.

1.2.2 Reasoning in the mathematics classroom

Mathematical reasoning associated with particular mathematical topics Important findings on mathematical reasoning by researchers in England is found in *Key Ideas in Teaching Mathematics, Research-based guidance for ages 9–19,* (Watson and Jones 2013). Four reasoning areas emerge from hard-to-learn and hard-to-teach topics: (1) ratio and proportional reasoning, (2) spatial and geometrical reasoning, (3) reasoning about data and (4) reasoning about uncertainty. In this section, we discuss the first two in detail.

Ratio and proportional reasoning The topic of ratio and proportional reasoning implies understanding concepts from a diverse set of approaches – such as those concerning fractions, ratios and scale factors – with contextual links. In the case of fractions, this means looking at division (an operation), a quotient (a result), a number and a part-whole relation. Ratios, on the other hand, could present either part-whole or part-part relations. Part-part relations lead to difficulties in choosing multiplicative strategies for comparing two ratios or finding a missing value. Scale factors contain the understanding of additive and multiplicative strategies and identifying the right one, especially when the scale factor is not integers. A linear thinking mode is a defining characteristic here.

When exploring geometry topics, a pathway opens up for bridging deductive reasoning and the abstraction of conceptual knowledge with its properties (Alghadari and Noor 2020). The notion of deduction serves as a valuable way to develop geometrical thought, as based on Van Hiele's five levels (from Level 0 to Level 4) of geometrical understanding, consisting of visualization, analysis, informal deduction (also known as abstraction), formal deduction and rigour (Shaughnessy and Burger 1985, p. 420). Burger and Shaughnessy (1986) refined these levels to include information on the predominant kinds of reasoning seen at each one:

1 Level 1: analysis – where students can contrast shapes while explicitly identifying relevant properties.
2 Level 2: information deduction/abstraction – where students interrelate shapes, making connections.
3 Level 3: formal deduction – where students frequently make conjectures and verify them. Here, information deduction/abstraction is one level below formal deduction, which leads to the question of how to nourish the abstraction process.

An example of Pythagorean theorem It is a classic topic in geometry and one which applies in other areas as well, where numerical computation and the relationship between geometrical properties come face to face. Alghadari and Noor (2020) used relationship-eliciting and relationship-integrating phases to describe its process of abstraction. The process consists of three elements: students' conceptual understanding, using visual representations to establish relationships and/or build rational rules, and logical argument to justify their purposive actions. How these three elements interact is also important. Among these three elements, visual representation is a way for students to manipulate geometrical properties to set up their algebraic thoughts so that they can argue visually (Fachrudin et al. 2019).

When looking into empirical research, a systematic review of relevant studies reveals that there are six domain-specific kinds of reasoning related to mathematical topics (Hjelte, Schindler and Nilsson 2020):

1 Spatial reasoning, that related to geometry.
2 Informational inferential reasoning, or reasoning with data.

3 Additive, multiplicative and distributive reasoning, where additive reasoning relates to part-whole relationships, multiplicative reasoning relates to ratios, and distributive reasoning relates to fractions.

4 Reasoning linking to algebra, going beyond arithmetic.

5 Proportional and covariational reasoning, where proportional relates to rates and covariational relates to ratios and rates.

6 Quantitative reasoning, that involving the exploration of relationships between quantities such as units of measurement, providing the foundation for algebraic reasoning.

When reasoning directs attention to particular topics as a way of achieving deeper understanding in those areas, it is important to note that reasoning forms not only a part of the learning journey, but also a destination, the goal for students to strive towards. That means we must recognize mathematical reasoning as something ideological, as Su (2017) pointed out, human beings long for truth via rigorous thinking, something cultivated by reasoning.

Mathematical reasoning not associated with topics After all of this discussion, then, what sense can we make from mathematical reasoning? One principal question that emerges is whether or not mathematical knowledge needs to be remembered and followed. Lithner (2008) split reasoning types into imitative reasoning and creative reasoning. Imitative reasoning is associated with rote learning. It also includes memorized reasoning, which heavily involves recollection of knowledge – for example, linear graphs in real life problems – and algorithmic reasoning, including explicit choice of a particular solution algorithm, for example solving an equation using a trial and improvement method. On the contrary, creative reasoning arises from non-routine tasks with elements of (1) novelty, i.e. constructing novel reasoning sequences, (2) plausibility, i.e. arguments for logical validity for the constructiveness, and (3) mathematical foundations. This dichotomy of using reasoning either known to students or new has formed the basis of empirical investigations (Hjelte, Schindler and Nilsson 2020). Based on Lithner's work, Adawiyah (2016) framed this as inductive-creative reasoning, noting three indicators: creative generalization, creative analogy and creative patterns.

Inductive, deductive and abductive Perhaps the most striking split in ideas of mathematical reasoning is that separating inductive and deductive kinds, as taken up in PISA. Another related type of reasoning is abductive reasoning, logical forms proposed by Peirce, Habermas and Eco (Reid 2018). This starts with an observation or set of observations and then seeks the simplest and most likely conclusion through the process of scientific exploration. Abductive reasoning brings an explanatory perspective to the reasoning process. If inductive reasoning leads to the establishment of reasonable hypotheses, and deductive reasoning ends up with logical conclusions, then abductive reasoning leads to plausible conclusions based on observation (Hidayah and Sa'dijah 2020).

Aliseda (2003, p. 26) explained the structure of abduction as 'deduction in reverse plus additional conditions'.

Adaptive reasoning Among the ideas research has proposed as the ultimate goal of mathematical learning is the notion of 'mathematical proficiency' (National Research Council 2001). This has five interwoven strands: adaptive reasoning, strategic competence, conceptual understanding, productive disposition, and procedural fluency. Adaptive reasoning means the 'capacity for logical thought and for reflection on, explanation of, and justification of mathematical arguments' (Kilpatrick 2001, p. 107). Adaptive reasoning subsumes deductive reasoning, which is based on the establishment of formal proof by following processes of logic and/or applying general rules to specific cases. It also encompasses inductive reasoning, which is based on observation and the interrogation of data on the way to making educated guesses.

Intuitive reasoning In addition to these two general reasoning types, intuition is also included in the adaptive reasoning process – in the forms of intuitive-inductive and intuitive-deductive reasoning (Muin, Hanifah and Diwidian 2018). Intuitive reasoning is based on the discovery of patterns or 'the patterning instinct', to borrow Jeremy Lent's book title (Lent 2017). Importantly, intuitive, inductive and deductive reasoning all overlap and need to be incorporated together in the development of student reasoning. One way to achieve this is the use of the creative problem solving (CPS) model. Muin, Hanifah and Diwidian (2018) have discovered that the CPS model supports intuitive-inductive aspects. Furthermore, Ansari, Taufiq and Saminan (2020) analyse Year 8 Indonesian students' adaptive reasoning with reference to five indicators: (1) the ability to propose a conjecture; (2) the ability to draw correct conclusions; (3) the ability to find the patterns of a problem; (4) the ability to present reasoning for the solution; and (5) the ability to examine the validity of an argument. The CPS model seems to impact on the first and last indicators most.

1.3 What comes next?

In this chapter, we have discussed the two basic dimensions of mathematical reasoning: reasoning in mathematics and reasoning as a mathematics learning goal in schools. Chapter 2 will turn to some of the foundational structures of proof that have shaped our understanding of reasoning in secondary settings.

Practice activities

Reflect on worked examples in your recent lessons: how are reasoning, understanding, problem-solving, and fluency linked to activities? Which one has been prioritized?

Box 1.3 Top tips on mathematical reasoning

1 Reasoning is needed even more than conceptual understanding and fluency; compared to reasoning, everything else in mathematics learning has only second-rate value.
2 Supporting students' understanding through connections; a logical connection between concepts requires time for the formation and elaboration of knowledge.

References

Adawiyah, R. (2016) *The effect of open approach towards mathematical inductive-creative reasoning ability.* Paper of Mathematics Education Department, Faculty of Tarbiyah and Teacher's training, Jakarta Syarif Hidayatullah State Islamic University, Jakarta, Indonesia.

Alderton, J., Donaldson, G., Ineson, G. et al. (2018) Pre-service primary teachers' approaches to mathematical generalization, *Proceedings of the British Society for Research into Learning Mathematics*, 37(3): 1–6. Available at: https://eprints.soton.ac.uk/418809/1/Approaches_to_math_generalisation_BSRLM_1117_pure.pdf (Accessed 6 September 2023).

Alghadari, F., and Noor, N.A. (2020) Students depend on the Pythagorean theorem: Analysis by the three parallel design of abstraction thinking problem, *Journal of Physics: Conference Series*, 1657(1): 012005. Available at: https://iopscience.iop.org/article/10.1088/1742-6596/1657/1/012005/pdf (Accessed 6 September 2023).

Aliseda, A. (2003) Mathematical reasoning vs. abductive reasoning: a structural approach, *Synthese*, 134(1/2), 25–44. http://www.jstor.org/stable/20117324

Ansari, B.I., Taufiq, T., and Saminan, S. (2020) The use of creative problem solving model to develop students' adaptive reasoning ability: inductive, deductive, and intuitive, *International Journal on Teaching and Learning Mathematics*, 3(1): 23–36. Available at: https://scholar.archive.org/work/ep52usovzbbohbcbytvtav4i3y/access/wayback/http://ejournal.uin-malang.ac.id/index.php/ijtlm/article/download/9439/pdf (Accessed 6 September 2023).

Bakker, A., Biehler, R., and Konold, C. (2004) Should young students learn about box plots?, *Curricular development in statistics education: International Association for Statistical Education*: 163–73. Available at: http://www.statlit.org/PDF/2004BakkerIASE.pdf (accessed 6 September 2023)

Ben-Zvi, D. (2004) Reasoning about data analysis, in D. Ben-Zvi and J. Garfield (eds) *The challenge of developing statistical literacy, reasoning and thinking.* Dordrecht, Germany: Springer. https://doi.org/10.1007/1-4020-2278-6_6

Biehler, R., Frischemeier, D., Reading, C., and Shaughnessy, J.M. (2018) Reasoning About Data, in D. Ben-Zvi, K. Makar and J. Garfield (eds) *International Handbook of Research in Statistics Education.* Springer International Handbooks of Education. Cham, Switzerland: Springer. https://doi.org/10.1007/978-3-319-66195-7_5

Blanton, M.L., Westbrook, S., and Carter, G. (2005) Using Valsiner's zone theory to interpret teaching practices in mathematics and science classrooms, *Journal of Mathematics Teacher Education*, 8(1): 5–33. https://doi.org/10.1007/s10857-005-0456-1

Brodie, K. (2009) *Teaching mathematical reasoning in secondary school classrooms* Vol. 775. Berlin/Heidelberg, Germany: Springer Science & Business Media. https://doi.org/10.1007/978-0-387-09742-8

Brodie, K. (2010) Pressing dilemmas: meaning-making and justification in mathematics teaching, *Journal of Curriculum Studies*, 42(1): 27–50. https://doi.org/10.1080/00220270903149873

Burger, W. F. & Shaughnessy, J. M. (1986) Characterizing the van Hiele levels of development in geometry, *Journal for research in mathematics education*, 17(1): 31–48.

Carraher, D.W., Martinez, M.V. and Schliemann, A.D. (2007) Early algebra and mathematical generalization, *ZDM Mathematics Education*, 40(1): 3–22. https://doi.org/10.1007/s11858-007-0067-7

Chen, C.L. and Herbst, P. (2013) The interplay among gestures, discourse, and diagrams in students' geometrical reasoning, *Educational Studies in Mathematics*, 83(2): 285–307. https://doi.org/10.1007/s10649-012-9454-2

Department for Education (2021) National curriculum in England: mathematics programmes of study. Available at: https://www.gov.uk/government/publications/national-curriculum-in-england-mathematics-programmes-of-study (accessed 16 June 2023).

Doerr, H.M. and English, L.D. (2003) A modeling perspective on students' mathematical reasoning about data, *Journal for research in mathematics education*, 34(2): 110–36. https://doi.org/10.2307/30034902

Eko, Y.S., Prabawanto, S. and Jupri, A. (2018) The role of writing justification in mathematics concept: The case of trigonometry, *Journal of Physics: Conference Series*, 1097(1):012146. Available at: https://iopscience.iop.org/article/10.1088/1742-6596/1097/1/012146/pdf (accessed 6 September 2023)

Ernest, P. (1991) *The philosophy of mathematics education*. London: Falmer Press.

Ernest, P. (1997) *The nature of mathematics and teaching*. Exeter, England: Perspectives-Exeter. [Online.]

Fachrudin, A.D., Ekawati, R., Kohar, A.W. et al. (2019) Ancient China history-based task to support students' geometrical reasoning and mathematical literacy in learning Pythagoras, *Journal of Physics: Conference Series*, 1417(1): 012042. Available at: https://iopscience.iop.org/article/10.1088/1742-6596/1417/1/012042/pdf (accessed 6 September 2023).

Goos, M. and Kaya, S. (2019) Understanding and promoting students' mathematical thinking: as review of research published in ESM, *Educational Studies in Mathematics*, 103: 7–25. https://doi.org/10.1007/s10649-019-09921-7

Gutiérrez S., Pearce D., Geraniou E., and Mavrikis M. (2008) Supporting Reasoning and Problem-Solving in Mathematical Generalization with Dependency Graphs, in G. Stapleton, J. Howse and J. Lee (eds) *Diagrammatic Representation and Inference, Diagrams 2008, Lecture notes in computer science, vol 5223*. Berlin/Heidelberg, Germany: Springer. https://doi.org/10.1007/978-3-540-87730-1_40

Herbst, P. (2004) Interactions with diagrams and the making of reasoned conjectures in geometry, *ZDM Mathematics Education*, 36(5): 129–39. https://doi.org/10.1007/BF02655665

Hidayah, I.N. and Sa'dijah, C. (2020) Characteristics of students' abductive reasoning in solving algebra problems, *Journal on Mathematics Education*, 11(3): 347–62. Available at: https://files.eric.ed.gov/fulltext/EJ1294576.pdf (accessed 6 September 2023).

Hjelte, A., Schindler, M. and Nilsson, P. (2020) Kinds of mathematical reasoning addressed in empirical research in mathematics education: a systematic review, *Education Sciences*, 10(10): 289. Available at: https://www.mdpi.com/2227-7102/10/10/289 (accessed 6 September 2023).

Kilpatrick, J. (2001) Understanding mathematical literacy: the contribution of research, *Educational Studies in Mathematics*, 47(1): 101–16. https://doi.org/10.1023/A:1017973827514

Lent, J. (2017) *The patterning instinct: a cultural history of humanity's search for meaning*. New York: Prometheus Books.

Lithner, J. (2008) A research framework for creative and imitative reasoning, *Educational studies in mathematics*, 67: 255–76. https://doi.org/10.1007/s10649-007-9104-2

Lo, J.J., Grant, T.J., and Flowers, J. (2007) Challenges in deepening prospective teachers' understanding of multiplication through justification, *Journal of Mathematics Teacher Education*, 11(1): 5–22. https://doi.org/10.1007/s10857-007-9056-6

Lowrie, T., Harris, D., Logan, T., and Hegarty, M. (2021) The impact of a spatial intervention program on students' spatial reasoning and mathematics performance, *The Journal of Experimental Education*, 89 (2): 259–77. https://doi.org/10.1080/00220973.2019.1684869

Martino, A.M. and Maher, C.A. (1999) Teacher questioning to promote justification and generalization in mathematics: What research practice has taught us, *The Journal of Mathematical Behavior*, 18(1): 53–78. https://doi.org/10.1016/S0732-3123(99)00017-6

McCluskey, C., Mulligan, J., and Mitchelmore, M. (2016) The Role of Reasoning in the Australian Curriculum: Mathematics, *Mathematics Education Research Group of Australasia*. Available at: https://files.eric.ed.gov/fulltext/ED572330.pdf (accessed 3 February 2023).

Mizrahi, M. (2020) Proof, explanation, and justification in mathematical practice, *Journal for General Philosophy of Science*, 51: 551–68. https://doi.org/10.1007/s10838-020-09521-7

Muin, A., Hanifah, S.H., and Diwidian, F. (2018) The effect of creative problem solving on students' mathematical adaptive reasoning, *Journal of Physics: Conference Series*, 948(1): 012001. Available at: https://iopscience.iop.org/article/10.1088/1742-6596/948/1/012001/pdf (accessed 6 September 2023).

Mulligan, J., Woolcott, G., Mitchelmore, M. et al. (2020) Evaluating the impact of a spatial reasoning mathematics program (SRMP) intervention in the primary school, *Mathematics Education Research Journal*, 32(2): 285–305. https://doi.org/10.1007/s13394-020-00324-z

National Research Council (2001) *Adding It Up: Helping Children Learn Mathematics*. Washington, D.C.: National Academy Press. https://doi.org/10.17226/9822

Nunes, T., Bryant, P., Evans, D. et al. (2015) *Teaching mathematical reasoning: probability and problem solving in primary school*. Department of Education, University of Oxford, England. Available at: https://nuffieldfoundation.org/sites/default/files/files/Nunes%26Bryant2015_Teachingreasoning%20-%2028Jan15.pdf (accessed 5 June 2023).

OECD (2018) *PISA 2021 Mathematics Framework (Draft)*. Available at: https://www.oecd.org/pisa/pisaproducts/pisa-2021-mathematics-framework-draft.pdf (accessed 2 July 2023).

OECD (n.d.) *PISA 2022 Mathematics Framework*. Available at: https://pisa2022-maths.oecd.org/ca/index.html#Mathematical-Reasoning (accessed 2 July 2023).

Reid, D.A. (2018) Abductive reasoning in mathematics education: approaches to and theorizations of a complex idea, *Eurasia Journal of Mathematics, Science and Technology Education*, 14(9): em1584. https://doi.org/10.29333/ejmste/92552

Shaughnessy, J.M. and Burger, W.F. (1985) Spadework prior to deduction in geometry, *The Mathematics Teacher*, 78(6): 419–28. https://doi.org/10.5951/MT.78.6.0419

Stylianides, G. (2008) An analytic framework of reasoning-and-proving, *For the Learning of Mathematics*, 28(1): 9–16. https://www.jstor.org/stable/40248592

Su, F.E. (2017) Mathematics for human flourishing, *The American Mathematical Monthly*, 124(6): 483–93. https://doi.org/10.4169/amer.math.monthly.124.6.483

TIMSS (2019) *TIMSS 2019 Mathematics Framework*. Available at: https://timss2019.org/wp-content/uploads/frameworks/T19-Assessment-Frameworks-Chapter-1.pdf (accessed 6 September 2023).

Wang, Y. (2015) *Understanding linear function in secondary school students: a comparative study between England and Shanghai.* [Doctoral thesis, Durham University]. ProQuest Dissertations and Theses Global. Available at: http://etheses.dur.ac.uk/11230 (accessed 6 September 2023).

Watson, A., Jones, K., and Pratt, D. (2013) *Key Ideas in Teaching Mathematics: Research-based guidance for ages 9-19.* Oxford, England: OUP.

Widjaja, W., Vale, C., Herbert, S. et al. (2020) Linking comparing and contrasting, generalizing and justifying: a case study of primary students' levels of justifying, *Mathematics Education Research Journal*, 33(2): 321–43. https://doi.org/10.1007/s13394-019-00306-w

Wilkerson, M.H. and Laina, V. (2018) Middle school students' reasoning about data and context through storytelling with repurposed local data, *ZDM Mathematics Education*, 50: 1223–35. https://doi.org/10.1007/s11858-018-0974-9

Wu, M. (2009) *A critical comparison of the contents of PISA and TIMSS mathematics assessments*, in NCES "What we can learn from PISA" research conference, Vol. 2, pp. 1–26. Available at: https://edsurveys.rti.org/PISA/documents/WuA_Critical_Comparison_of_the_Contents_of_PISA_and_TIMSS_psg_WU_06.1.pdf (accessed 18 June 2023).

Zayyadi, M. and Kurniati, D. (2018) Mathematics reasoning and proving of students in generalizing the pattern, *International Journal of Engineering & Technology*, 7(2): 15–7. Available at: https://www.researchgate.net/profile/Dian-Kurniati-2/publication/324527323_Mathematics_reasoning_and_proving_of_students_in_generalizing_the_pattern/links/5af79aaba6fdcc0c031f2a0e/Mathematics-reasoning-and-proving-of-students-in-generalizing-the-pattern.pdf (accessed 6 September 2023)

2 Reasoning and proof in formulas

Key arguments

In the second chapter, we argue that proof is the basic building block of logical discourse, forming the very foundation of structural continuities within the web of mathematics knowledge. Written proof, for example, is the source for the internal development of subject knowledge in teachers as well as students.

Key terms

- Written proof

  The process of moving from mathematical statements to conclusions is proof, and often this takes written form. Written proof reveals underlying logic by employing a two-column format: the left side on the page is the statement, and the right side gives the reasons.

- The relationship between justification and proof

  Chapter 1 discussed *justification by mathematizing* and *proof by conjecturing*, and in this chapter we turn to clarifying the boundary between justification and proof. Justification is launched by proof, but proof does not necessarily depend on justification. Proof has rigour and formality, with the purpose of convincing oneself as well as others.

In the first part of this chapter, we aim to give you a new way to think about a specific notion related closely to mathematical reasoning, proof. We consider what it is, how to achieve it, and how it features in the curricula of different countries. Then, we zoom into the area of geometry, which is where proof appears most often in curricula internationally. Geometry involves various 'how' inquiries: how we can organize the web of knowledge, and how we can develop intersections between the graphic representation of geometric elements and the conceptual understanding of their meaning, in order to form a written proof. We take two examples from geometry to introduce the formal proof style: trigonometry formulas and circle theorems. This practical part of the chapter is aimed at subject-knowledge enhancement for those who are interested in a formal style of step-by-step deductive proof. You are welcome to pick one or two sections that would be mostly relevant to your interests.

2.1 Proof and proving

A colleague, John, loves researching his family history. He announced that he might be a descendent of King James I on a day when we happened to be sitting in the King James I Academy in Bishop Auckland, Durham, for a professional development day. We teasingly addressed him as 'Your Royal Highness', but John replied that any of us could be related to the royal family and here is his 'proof': if you go back one generation, then two people are related to you, expressed as 2^1. If you go back two generations, then four people are related to you, 2^2. Going back long enough, the number of people related to a person alive today will account for the whole population at that given time. Doing some quick maths, $2^{21} = 2,097,152$, $2^{22} = 4,194,304$, England's population in early 1600s is around 3.73 million. So, a person alive today could theoretically be related to most of the people living then.

This is an exact example of generalization by explanation, starting from the specific. We could classify it as empirical argumentation, or empirical exploration. But does this deserve to be called a proof? Suppose the further question is what the unique characteristics of proof are in school mathematics. In this section, we will discuss three related questions: (1) how is proof conceptualized?; (2) how is proof achieved and what types of proof are there?; and (3) how does proof appear in the curricula of different countries?

2.1.1 What is proof?

In chapter 1, we summarized three types of reasoning: generalization by explanation, proof by conjecturing, and justification by mathematizing. The first type has also been referred to as reasoning of inquiry, while the other two have been called reasoning of justification (Ball et al. 2002). But is empirical argumentation, like John's claim about possible royal descent, an example of proof? Stylianides (2007) excluded empirical arguments from being considered proofs mainly because of invalid modes of argumentation based on incomplete evidence. He also described proof in school mathematics classes as 'mathematical argument' with three dimensions: (1) accepted statements; (2) modes of argumentation, such as direct proof, and the use of counter-examples; and (3) modes of representing arguments, for example using language, pictorial and abstract representation. This leaves two further questions, concerning the differences between justification and proof, and the differences between argumentation and proof.

The differences between justification and proof This is about reasoning of justification. Bergwall (2021a) explained justification as stemming from the purpose of convincing, of persuading others of the truth of a mathematical statement. Proof, on the other hand, is geared towards evaluating justifications according to how rigorous, precise and formal they are. Thus, Bergwall calls proof 'a connected sequence of assertations' (2021a, p. 735), and this sequence is then followed by justification. Justification, therefore, is launched by proof, but proof does not depend on justification. The essential point is to link the concept of proof with rigour and formality.

The differences between argumentation and proof Pedemonte (2007) acknowledged that both argumentation and proof involve rational justification, and both have the purpose of convincing a universal audience of the truth of mathematics statements in a theoretical field, such as algebra. Toulmin (1993) characterized the process of argumentation as moving from data to claim through a warrant. The warrant could be a theorem, definition or an axiom to connect the data and claim. From the point of view of Stylianides' definition, proof is a special kind of argumentation. But how special is it?

Pedemonte (2007) took a structural perspective – highlighting deductive, inductive and abductive kinds – to analyse the differences, especially structural continuities within cognitive unity, such as algebra, and structure distances. Findings suggest that one of the difficulties that students meet concerns transforming the structure of argumentation into a deductive type in a proof. In the area of algebra, Pedemonte (2008) revealed how, in numbers, an abductive structure of argumentation facilitates the deductive structure of proof. What then about the link between arithmetic and algebra? Martinez and Pedemonte (2014) later looked at an example of inductive arithmetical argumentation and algebraic deduction. Their findings illustrated two differences: that argumentation depends on arithmetic while proof requires algebra, and that argumentation has an inductive nature whereas proof has a deductive nature. This highlights the concept of proof as something based on deductive reasoning.

To summarize the answers to the two questions, proof feeds into justification and argumentation, and it is the end product of a series of interactions with argumentation and justification. Within the school mathematics context (as opposed to in the context of a mathematician's proof), the focus is the logical side of proof. But the social nature of proof and its practical value, which are actually behind the end product of proof, cannot be underestimated. Proof or the construction of proof, then, not only illuminates why a mathematical claim is correct, but it is also about finding the most convincing way possible for others to embrace the mathematical claim. In this sense, the concept of proof comes down to 'demonstration' – according to Balacheff (1988) and Miyazaki (2000). Proof, therefore, has a social interactive meaning. A proof scheme has the two sides of convincing yourself and convincing someone else. This points to the subjective side of proof.

Box 2.1 Proof

1 Justification is launched by proof, but proof does not depend on justification.
2 Proof has rigour and formality to convince others as well as yourself.

The vision of proof elaborated so far highlights its deductive nature and its rigour and formal style but also embraces aspects of social interaction. These former aspects imply decontextualization of representation thereby promoting abstraction. A didactical sense of proof measures rigour from two perspectives: (1) being geared to the internal development of mathematics rather than being application oriented, i.e. the proof itself and (2) involving well-defined objectives rather than experience-based argument (Balacheff, 2010) i.e. a requirement/convention to write proof in a certain way.

In addition to these considerations, there are levels of proof just like there are levels of understanding. For instance, there is the active process of finding a proof, or proving, which ultimately facilitates learning. We will explore this next, starting from the question of how to actively engage in proving.

2.1.2 How to prove

Balacheff (1988) argued for the importance of cognitive and linguistic bases in the process of proving, distinguishing between two types of proof: pragmatic proof and intellectual proof. Pragmatic proof uses an experimental approach, mainly in the form of verbal formulation, in which students' mathematics conceptions are not theorized. Hanna (2000) countered that the explorative nature of this experimental approach does not count as proof at all. Intellectual proof, on the other hand, can involve several different processes. These are valued hierarchically, moving upwards from:

1 Naïve empiricism, mainly via observation, in which students do not try to enter a proving process by themselves, or in which students check their solution against given examples or documents.
2 Crucial experiment, which reveals the awareness to validate a mathematical statement, with element of generalization.
3 Thought experiment, relying on a generic example via a de-contextualization process, which is a lower level of proving.
4 Treatment of refutation, how to treat counter-examples, reject them, considering them as exceptions or as reason to introduce a condition.

Balacheff's work has ultimately convinced the mathematics education community that the foundation of proving, or teaching proof, lies in the role of contradiction, which bridges between students' empirical experience and deductive explanations. In the case of geometrical proof, one possible solution is to use dynamic geometry software (Jones 2000; Hadas, Hershkowitz and Schwartz 2000).

Based on Balacheff's work, Miyazaki (2000) proposed six levels of proof at lower-secondary level within the area of algebra, focusing on three dimensions: (1) contents, whether inductive or deductive reasoning; (2) representations, whether functional language of demonstration or others; and (3) students' thinking, whether involving formal or concrete operations. Among these three dimensions, the most advanced level has deductive reasoning at its core, in tandem with the functional language of demonstration via formal operations. The least advanced level is inductive reasoning with non-functional language via concrete operations. Köğce, Aydın and Yıldız (2010) used Miyazaki's levels to examine higher-secondary-level students' views about proof. The majority of these students reported that their confidence in mathematics was established by engaging in proving, as their desire was to learn 'the trueness of a proposition with its reasons' (Köğce, Aydın & Yıldız, 2010, p. 2548).

These findings have, in some ways, transformed understandings of proof; proof becomes a means to engage students with mathematics learning. That is not to say, however, that this has necessarily been fully absorbed in national curricula. So, the next section will investigate curricula in different nations and textbooks, considering how proof is promoted, how students are expected to go about proving, at what levels and of what types.

2.1.3 Proof in national curricula and textbooks

National curricula not only provide subject contents, but they also indicate the competencies that students should hold in dealing with these contents. Research on proof-related competences reveals ways in which students in particular places make sense of proof.

The content of proof Hemmi, Lepik and Viholainen (2013) used six categories to compare mathematics curricula in Estonia, Finland and Sweden: (1) proof, mainly deductive proof; (2) argumentations; (3) investigations; (4) structures; (5) definitions; and (6) logic. Let's take the lower-secondary level as the comparative example. The Estonian curriculum has 'a serious focus' on proof in the area of geometry, including on Pythagoras and circle theorems (see section 2.3). Finland, on the other hand, stresses mathematical structure and the systematic trial-and-error method, but this is not limited to the area of geometry. The Swedish curriculum does not mention proof, but instead focuses on argumentation – for instance, arguing for validity. These distinct differences in national curricula might reflect how each country balances the two sides of proof: a rigorous and formal style via deductive reasoning versus a demonstrative style via convincing oneself and others.

The progress of proof If the discussion above concerns the outputs of proof, then it raises the question of what the progress looks like. Rind and Mughal (2020) criticized that the fact that although reasoning and justification skills are listed among the five standards in Pakistan's national curriculum for mathematics, the curriculum does not provide a roadmap for developing them. Millar (2011) called for clear communication of intended curriculum content through examples, or through explanation of key concepts at each stage, as a means of clarifying what progression in this area looks like. Similar was found in England by Oates (2011), who recommended two focuses for education reform: developing curriculum coherence between the central national curriculum and local scheme of works in schools; and curriculum control to maintain closes coherence at different levels.

Popular topics related to proof

One possible way to pursue the intention of building coherence is to use textbooks, which serve to exemplify expectations for implementation. Fan and his colleagues (Fan et al. 2018) compared examples on geometric proof from textbooks across countries where compulsory textbooks are used: China, Indonesia and Saudi Arabia. These examples were categorized into the three types of proof that were featured: direct proof, proof by contradiction and counter-examples. The work revealed that none of the textbooks use counter-examples, but instead that they place emphasis on the process of deduction, such as in direct proof. The Chinese textbooks cover direct proof in a range of topics, such as parallel lines, triangles, circles, and parallelograms, while in Indonesia's and Saudi Arabia's textbooks, most examples are presented within the topic of triangles, e.g. congruent triangles.

Social sides of proof If this work to categorize examples reflects the deductive nature of proof, then the social sides of proof were investigated by Bergwall (2021b) through the identification of two aspects to proof-related worked examples: (1) developing or investigating a statement, and (2) developing or investigating an argument. When comparing worked examples in integral calculus from Finland and Sweden, it was found that developing an argument is the predominant focus in Finnish textbooks. Therefore, at the implementation stage, empirical proof schemes (i.e. using examples) and deductive proof schemes are emphasized and promoted in different countries. This, however, has not prevented the deductive nature of proof from being broadly accepted.

Research has consistently illustrated that students in secondary settings who may be good at producing oral explanations often struggle to incorporate this into written proofs (Winer and Battista 2022; Soto-Johnson and Fuller 2012; Stylianides 2019). Therefore, in the next section, we will turn particular attention to written proof.

2.1.4 Written proof in geometry

Although the exact nature of proof is defined by each national curriculum (Miyakawa 2012), students across the world have been found to share problems and difficulties with written proof, especially in geometry, which is an area most likely to see common requirements across national curricula. The problem lies not with proving per se, but rather with the writing element. The two-column proof (or the flow-proof format, advocated by McMurray (1978) and Basinger (1979), has become deeply rooted in the mathematics community going back to the beginning of the twentieth century (Herbst 2002). 'Two-column' refers to the format of a statement on the left side and an explicitly stated reason on the right side, e.g. axioms, theorems, etc.. Through the specific order of these statements, proof becomes an exercise in logic, of showing how propositions – the givens – lead to conclusions.

Here is the simplest example of the two-column method, used to find the third angle in a given triangle:

<u>Given</u>: If in Triangle ABC, angle A (referred to as $\angle A$) is 60°
(i.e. $\angle A = 60°$) and $\angle B = 70°$, then what is the size of $\angle C$?

<u>Argument</u>
$\angle A + \angle B + \angle C = 180°$ (the sum of angles of a triangle is 180°)
$\angle A = 60°$; $\angle B = 70°$ (given)
So $60° + 70° + \angle C = 180°$
$\angle C = 50°$

So, which one is more important for students: the geometric truth, the logic, or the format of the logic? As this particular format deals in writing (something that can have various different styles), there is a danger that an exercise in the logical process (how to order these statements with corresponding reasons) becomes an exercise in memorizing the logically reasoned text. But the main point is the logic, how arguments are organized to express the logic.

In the following sections, we take two examples from geometry to demonstrate this format for a proof:

1 Three trigonometry formulas
   Using the basic trigonometry to explore the area of a triangle, using the area of a triangle to prove the sine rule, and using the unit of the circle to prove the cosine rule.
2 Eight circle theorems
   Using the properties of isosceles triangles to prove the eight circle theorems.

2.2 Proof for three trigonometry formulas

To make the proof concise and easy to understand, we will use the symbol 'Δ' to stand for a triangle; for example, triangle ABC is written ΔABC.

2.2.1 Sine formula for area of a triangle

Diagram 2.1

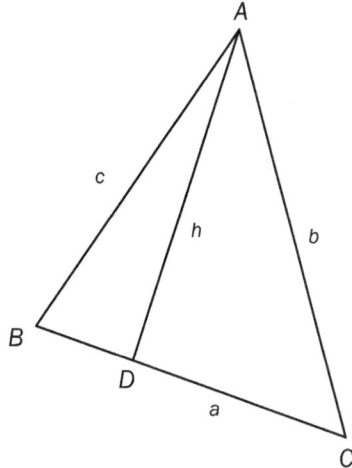

> **Area of △ABC (see Diagram 2.1):**
>
> $\frac{1}{2} ab\,SinC = \frac{1}{2} bc\,SinA = \frac{1}{2} ac\,SinB$

Outline of proof The proof involves using the standard formula for the area of a triangle; you will be familiar with this as: Area of $\triangle ABC = \frac{1}{2} ah$

    With a being the base upon which h, the perpendicular height, sits. All that we will be doing is finding a different expression for h and substituting this in. So:

Formal proof The formula for the area of $\triangle ABC = \frac{1}{2} ab$ sin$C$ or $\frac{1}{2} bc$ sin$A$ or $\frac{1}{2} ac$ sin$B$

<u>Argument</u>

sin$B = \frac{h}{c}$ (basic definition of sine ratio for $\angle B$ at Rt.$\triangle ADB$)

So $h = c \times$ sin$B$ (via rearrangement)

Area of $\triangle ABC = \frac{1}{2} ah = \frac{1}{2} a \times c \times$ sin$B$ (via substitution)

    The same proof process can be applied to the other two formulas for the area of a triangle.

2.2.2 Sine rule

Diagram 2.2

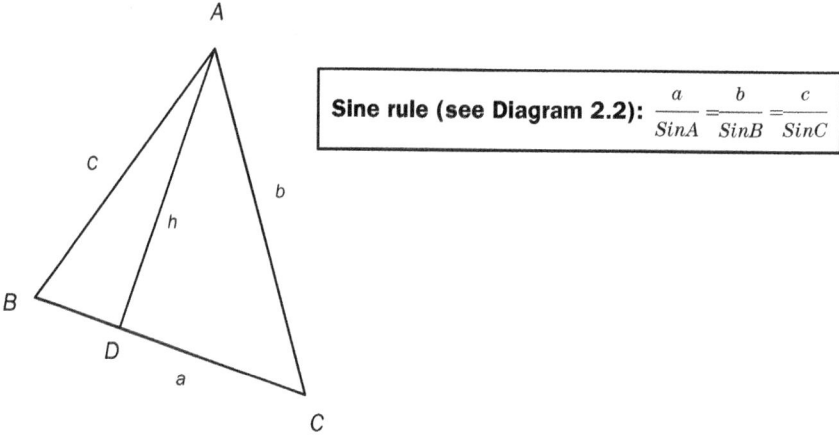

Outline of proof We will show two different variations. For the first we will use the sine area formulas that we have just developed and equate them for a simple proof.

Secondly, we will define h via basic trig ratios and just equate these.

Note how these two proofs are effectively the same since we have used the definition of h to derive the area formulas.

Formal proof 1: Argument from the perspective of the area of a triangle

Proof for the sine rule:

$$\frac{a}{\sin A} = \frac{b}{\sin B} = \frac{c}{\sin C}$$

Argument

Area of $\triangle ABC = \frac{1}{2} ab \sin C = \frac{1}{2} bc \sin A$ (selecting two of our area formulas)

So $\frac{1}{2} ab \sin C = \frac{1}{2} bc \sin A$

$\quad ab \sin C = bc \sin A$ (multiply both sides by 2)

$\quad a \sin C = c \sin A$ (divide both sides by b)

$\quad \frac{a \sin C}{\sin A} = c$ (divide both sides by $\sin A$)

$\quad \frac{a}{\sin A} = \frac{c}{\sin C}$ (divide both sides by $\sin C$)

Use the same process to prove $\frac{a}{\sin A} = \frac{b}{\sin B}$

Formal proof 2: Argument from basic trig ratios

Proof for the sine rule: $\frac{a}{\sin A} = \frac{b}{\sin B} = \frac{c}{\sin C}$

Argument

$$\sin B = \tfrac{h}{c} \qquad \text{(basic definition of sine ratio for } \angle B \text{ at Rt.}\triangle ABD)$$

So $h = c \times \sin B$ (rearrange)

$$\sin C = \tfrac{h}{b} \qquad \text{(basic definition of sine ratio for } \angle C \text{ at Rt.}\triangle ACD)$$

So $h = b \times \sin C$ (rearrange)

Then $c \times \sin B = b \times \sin C$ (equate both equations, $h = c \times \sin B$ and $h = c \times \sin B$)

$$\tfrac{c \times \sin B}{\sin C} = b \qquad \text{(divide both sides by } \sin C)$$

$$\tfrac{c}{\sin C} = \tfrac{b}{\sin B} \qquad \text{(divide both sides by } \sin B)$$

Use the same process to prove $\tfrac{a}{\sin A} = \tfrac{b}{\sin B}$

2.2.3 Cosine rule

Diagram 2.3

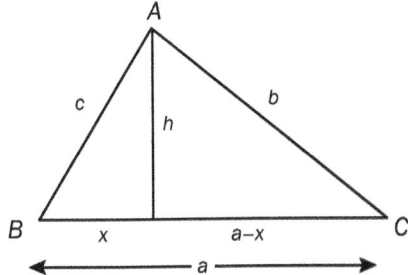

Cosine rule (see Diagram 2.3):

$$a^2 = b^2 + c^2 - 2cb \times CosA$$

$$b^2 = a^2 + c^2 - 2ac \times CosB$$

$$c^2 = a^2 + b^2 - 2ab \times CosC$$

Outline of proof Utilizing the diagram (where we have split the triangle into two right-angled triangles), we will use Pythagoras' theorem to construct a basic equation for each right-angled triangle and then, since 'h^2' will be a common term, use this to construct a further equation, after which we will substitute in a basic trig ratio.

Formal proof Proof for cosine rule: $b^2 = a^2 + c^2 - 2ac \times \cos B$

Argument

$c^2 = x^2 + h^2$ (Pyth. th. for left hand side right-angled triangle)

So $h^2 = c^2 - x^2$ (re-arrange with h^2 as subject)

$b^2 = (a - x)^2 + h^2$ (Pyth. th. for right hand side right-angled triangle)

So $h^2 = b^2 - (a - x)^2$ (re-arrange with h^2 as subject)

$c^2 - x^2 = b^2 - (a - x)^2$ (equate both: $h^2 = c^2 - x^2$ and $h^2 = b^2 - (a - x)^2$)

$c^2 - x^2 = b^2 - (a^2 + x^2 - 2ax)$ (expanding the brackets)

$c^2 - x^2 = b^2 - a^2 - x^2 + 2ax$ (multiplying the sign through)

$c^2 = b^2 - a^2 + 2ax$ (cancelled x^2 on both sides)

$\cos B = \tfrac{x}{c}$, so $x = c \times \cos B$

Since
$$c^2 = b^2 - a^2 + 2ax$$
$$c^2 = b^2 - a^2 + 2a \times c \times \cos B \quad \text{(substitute in for } x\text{)}$$
$$-b^2 = -a^2 - c^2 + 2ac \times \cos B \quad \text{(rearrange)}$$
$$b^2 = a^2 + c^2 - 2ac \times \cos B \quad \text{(multiply both sides times by } -1\text{)}$$

Use the same process to prove the other two formulas:
$$a^2 = b^2 + c^2 - 2bc \times \cos A$$
$$c^2 = a^2 + b^2 - 2ab \times \cos C$$

2.3 Proof of eight circle theorems

To make the proofs concise and easy to understand, we will use the symbol '$\angle$' to stand for angles; for example, angle C is written as $\angle OCA$ (see Diagram 2.4).

Diagram 2.4

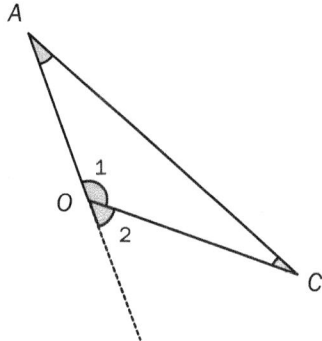

We assume basic triangle facts are known and understood, i.e.
$$\angle OAC + \angle OCA + \angle 1 = 180° \quad \text{(the sum of interior angles is 180°)}$$
$$\angle 1 + \angle 2 = 180° \quad \text{(because angles on a straight line add up to 180°)}$$
$$\angle OAC + \angle OCA = \angle 2 \quad \text{(an exterior angle equals the sum of the two other interior angles)}$$
Also, if $\triangle OAC$ is an isosceles triangle, $OA = OC$, $\angle OCA = \angle OAC$, then $2\angle OCA = \angle 2$ (see Diagram 2.4)

2.3.1 Circle theorem 1

The angle subtended by an arc at the centre is twice the angle subtended at the circumference.
$\angle BOC$ (the centre of the circle) is double $\angle BAC$ (which lies on the circumference).
So $\angle BOC = 2\angle BAC$ (see Diagram 2.5)

Diagram 2.5

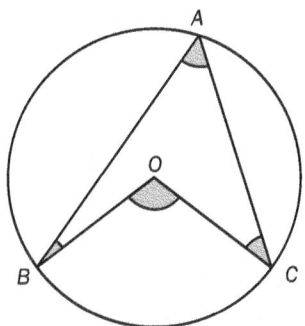

Outline of proof Split the 'arrowhead' shape into two isosceles triangles (see Diagram 2.6) and use our basic angle facts to prove the statement.

Formal proof
<u>Argument</u>

Diagram 2.6

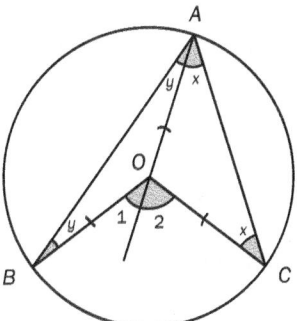

On the diagram, construct the line AO as an extended radius. (This divides the arrowhead shape into two isosceles triangles) and the central highlighted angle is split into $\angle 1$ and $\angle 2$ (See Diagram 2.6).
Since for $\triangle OAC$:
$OA = OC$ (both are radii)
therefore $\angle OCA = \angle OAC = x$
Base angles are now labelled as 'x' and with identical argument for $\triangle OAB$ its base angles are labelled with 'y'.

$\angle 2 = 2\angle OCA = 2x$ (exterior angles equals the sum of the interior opposite angles)
$\angle 1 = 2\angle OBA = 2y$ (exterior angles equals the sum of the interior opposite angles)

Therefore

$\angle 1 + \angle 2 = 2x + 2y = 2(x + y)$
$\angle BOC = 2\angle BAC$

Note: Be aware of the conceptual variation of this theorem (see Diagram 2.7)

Diagram 2.7

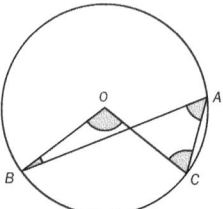

Note: Diagrams illustrating circle theorem 1 typically position the chord below the centre of the circle and the angle(s) subtended at the circumference above it (see Diagrams 2.5 and 2.7). In Diagram 2.8, the same rule applies to the angles AOC – the *reflex* angle at the centre – and ABC – the angle subtended at the circumference – where both the chord and the angle at the circumference are below the centre of the circle. Be aware of this conceptual variation.

Diagram 2.8

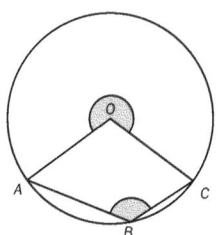

2.3.2 Circle theorem 2

Angles at the circumference subtended by the same arc (or the same segment) are equal.

So, $\angle BAC = \angle BDC$ (see Diagram 2.9).

Diagram 2.9

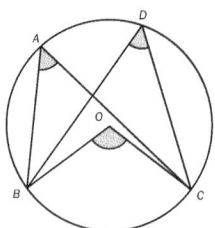

Outline of proof This follows naturally from circle theorem 1.

Formal proof

Argument

$\angle BOC = 2\angle BAC$ (circle theorem 1)

$\angle BOC = 2\angle BDC$ (circle theorem 1)

Therefore

$2\angle BAC = 2\angle BDC$ (equating via $\angle BOC$)

$\angle BAC = \angle BDC$ (halve both sides)

2.3.3 Circle theorem 3

Opposite angles in a cyclic quadrilateral sum to 180°.
So $\angle BAD + \angle BCD = 180°$, $\angle ABC + \angle ADC = 180°$ (see Diagram 2.10)

Diagram 2.10

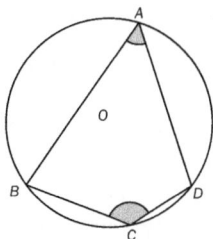

Outline of proof Via the construction of two radii, we will divide the cyclic quadrilateral into two pieces (see Diagram 2.11) which will conform with the typical diagram of circle theorem 1. By applying circle theorem 1 to this, we will prove $\angle BAD + \angle BCD = 180°$, $\angle ABC + \angle ADC = 180°$

Formal proof
<u>Argument</u>
Diagram 2.11

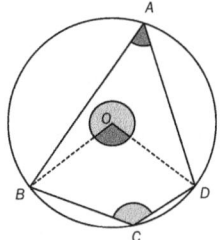

$$\angle BOD = 2\angle BAD \quad \text{(circle theorem 1)}$$
$$\text{Reflex } \angle BOD = 2\angle BCD \quad \text{(circle theorem 1)}$$
$$\angle BOD + \text{reflex } \angle BOD = 360° \quad \text{(angles around a point add up to 360°)}$$
$$\text{So } 2\angle BAD + 2\angle BCD = 360°$$
$$\angle BAD + \angle BCD = 180°$$
The same process applies to $\angle ABC + \angle ADC = 180°$.
Thus opposite angles in a cyclic quadrilateral sum to 180°.
    Note: Be aware of conceptual variation of this theorem (see Diagram 2.12).

Diagram 2.12

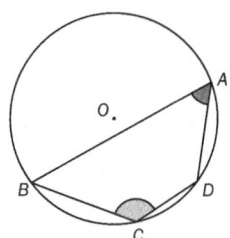

2.3.4 Circle theorem 4

The angle at the circumference in a semi-circle is a right angle.
So $\angle BAC = 90°$ (see Diagram 2.13)

Diagram 2.13

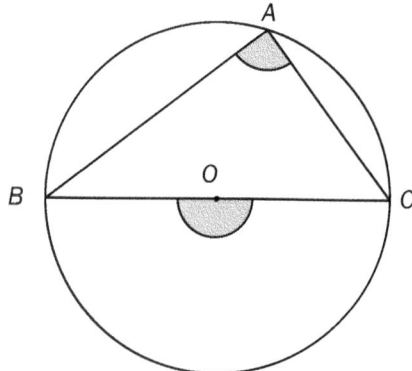

Outline of proof This is simply a specific instance of circle theorem 1

Formal proof
Argument

$\angle BOC = 180°$	(the angle on a straight line is 180°)
$\angle BOC = 2\angle BAC$	(circle theorem 1)
Therefore $2\angle BAC = 180°$	(equating $\angle BOC$)
$\angle BAC = 90°$	(half both sides)

2.3.5 Circle theorem 5

The angle between a tangent and a radius is 90°.
This is illustrated in Diagram 2.14.

Diagram 2.14

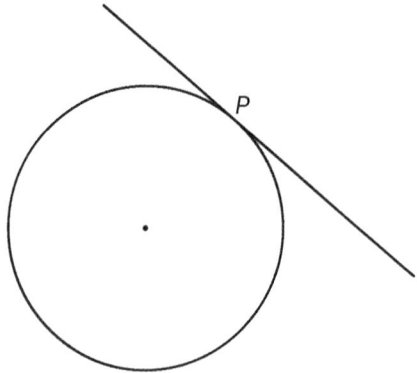

Outline of proof From arguments of symmetry, the basic property of a line touching a circle at a single point defines tangency.

Formal proof A circle can be considered to have infinite lines of symmetry. Similarly, an infinite line can be considered to have an infinite number of points through which a line perpendicular to that line can be drawn, thereby producing a single line of symmetry.

A line touching a circle at point P will always therefore produce a diagram with a single line of symmetry (see Diagram 2.15).

Since the angle on a straight line is 180° then, via symmetry, the angle formed between a 'tangent' line and a radius must be 90°.

Diagram 2.15

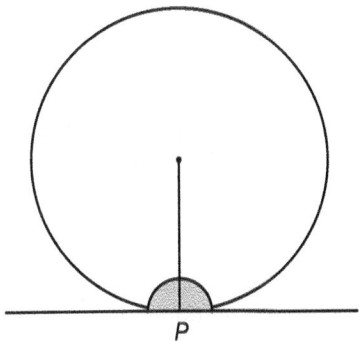

P

2.3.6 Circle theorem 6

The perpendicular from the centre of a circle to a chord bisects the chord.
If $OA \perp BC$ in the circle with centre O, then $BD = DC$ (Diagram 2.16).

Diagram 2.16

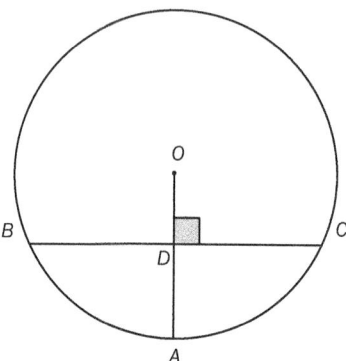

Outline of proof Given a perpendicular line OA cutting a chord BC, we will use the construction of isosceles triangles and the properties of congruency to prove that $BD = DC$.

Formal proof
Argument
Construct lines OB and OC (see Diagram 2.17)
$OA \perp BC$, then $\angle ODC = \angle ODB = 90°$
$\triangle ODC$ and $\triangle ODB$ are both right-angled triangles
$OD = OD$ (the shared side for both right-angled triangles)
$OB = OC$ (both lines are radii of the circle)
So Rt.$\triangle ODC$ and Rt.$\triangle ODB$ are congruent.

Therefore
$BD = DC$ (the corresponding sides are the same in the congruent triangles)
D is the midpoint of BC.

Diagram 2.17

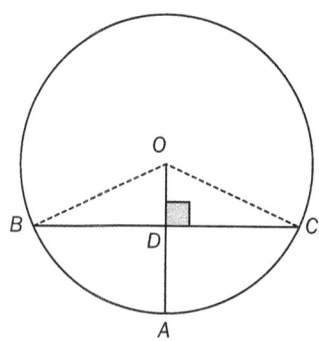

Converse of circle theorem 6: A line bisecting a chord and passing through the centre of the chord's circle must be perpendicular to the chord.

If $BD = DC$ by radius OA through point D, then $OA \perp BC$ (see Diagram 2.18).

Diagram 2.18

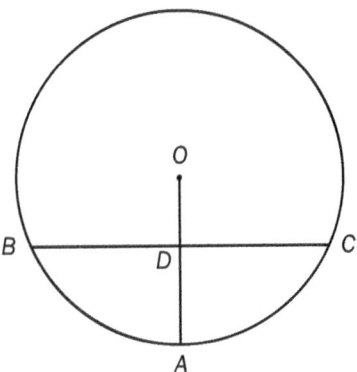

Outline of proof Given a chord BC bisected via a radius OA, we will again use congruency to prove that $\angle ODB$ is 90° and therefore $OA \perp BC$.

Formal proof

Diagram 2.19

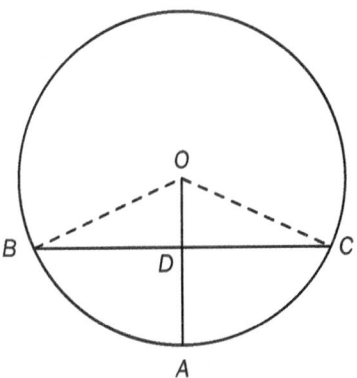

<u>Argument</u>
Construct lines OB and OC (see Diagram 2.19).
Within $\triangle ODC$ and $\triangle ODB$
$OD = OD$ (the shared side for both triangles)
$OB = OC$ (both lines are radii of the circle)
$BD = DC$ (D is the midpoint of BC)
So $\triangle ODC$ and $\triangle ODB$ are congruent   (S.S.S.)
        $\angle ODC = \angle ODB$ (corresponding angles are the same in congruent
                triangles)
$\angle ODC + \angle ODB = 180°$ (the angle on a straight line is 180°)
Then $\angle ODC = \angle ODB = 90°$, so $OA \perp BC$

2.3.7 Circle theorem 7

Two tangents drawn from the same point to a circle are equal in length.
So $AB = AC$

Diagram 2.20

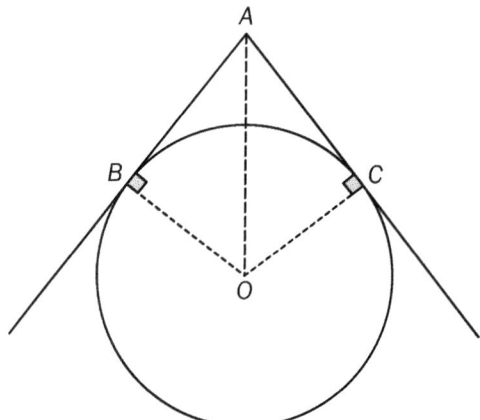

Outline of proof We will continue to use the principles of congruency in right-angled triangles to prove that $AB = AC$

Formal proof

<u>Argument</u>

Construct the lines OB, OC and OA (see Diagram 2.20)

$\angle ACO = 90°$ (AC is a tangent), then $\triangle OCA$ is a Rt.$\triangle$

$\angle ABO = 90°$ (AB is a tangent), then $\triangle OBA$ is a Rt.$\triangle$

With Rt. $\triangle OCA$ and Rt. $\triangle OBA$,

$OC = OB$ (both lines are radii of the circle)

$AO = AO$ (the shared side for both triangles)

So Rt.$\triangle OCA$ and Rt.$\triangle OBA$ are congruent.   (RHS congruency rule for right-angled triangles)

$AC = AB$ (corresponding sides are the same in congruent triangles)

2.3.8 Circle theorem 8

The angle between a tangent and a chord is equal to the angle in the alternate segment.
So $\angle BCD = \angle CAB$ (see Diagram 2.21)

Diagram 2.21

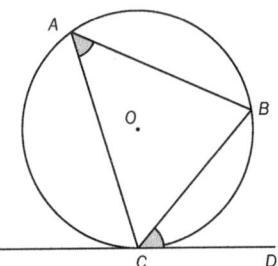

Outline of proof By constructing an isosceles triangle and using the facts associated with this, along with Circle Theorem 1, we will prove that $\angle BCD = \angle CAB$

Formal proof

Diagram 2.22

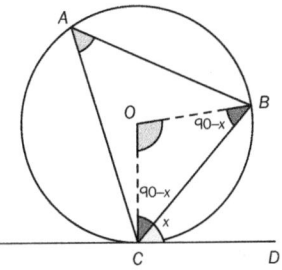

Argument
Construct lines OC and OB thereby creating an isosceles triangle (see Diagram 2.22)

$$\angle OCD = 90° \qquad \text{(circle theorem 5)}$$

Assume $\angle BCD = x°$, then $\angle OCB = 90° - x°$

$$OC = OB \qquad \text{(both are radii of the circle)}$$

So $\angle OCB = \angle OBC$ (property of isosceles triangles)

$$\angle OCB = \angle OBC = 90° - x°$$

and

$$\angle OCB + \angle OBC + \angle COB = 180° \quad \text{(the sum of interior angles in a triangle is 180°)}$$

therefore

$$90° - x° + 90° - x° + \angle COB = 180°$$

$$\angle COB = 180° - (90° - x° + 90° - x°) \text{ (rearrange)}$$

$$\angle COB = 2x° \qquad\qquad\qquad \text{(simplify)}$$

$$\angle COB = 2\angle CAB = 2x° \qquad \text{(circle theorem 1)}$$

$$\text{Then } \angle CAB = x°$$

$$\angle BCD = x°$$

$$\text{So } \angle BCD = \angle CAB$$

2.4 What comes next?

This chapter summarizes the nature of proof as a manifestation of deductive reasoning with its own rigours and formality.

Box 2.2 Top tips on proof

1 Proof is internally oriented towards the connections between mathematics topics rather than application oriented.
2 There are two broad types: empirical experience and deductive explanations. We need a balance between both types in school maths, taking care not to overlook the latter.
3 Sine and cosine rules can be deduced from Pythagoras and basic trigonometry.
4 All circle theorems can be deduced from circle theorem 1.

The next part will turn to lesson planning, in particular taking a mastery approach to lesson planning, which we understand as hypothetical teaching and learning trajectories. We will examine how to embed reasoning into lesson planning.

Practice activities

Circle theorem 1 states that the angle subtended by an arc at the centre is twice the angle subtended at the circumference. How would you prove that this is true? You first have to convince yourself. Then, imagine that you have a penfriend; how could you convince your friend in writing?

References

Balacheff, N. (1988) *A study of students' proving processes at the junior high school level.* Paper presented at the 66th Annual Meeting of the National Council of Teachers of Mathematics, Chicago, USA.

Balacheff, N. (2010) Bridging knowing and proving in mathematics: a didactical perspective, in G. Hanna, H.N. Jahnke and H. Pulte (eds) *Explanation and proof in mathematics—philosophical and educational perspectives*, pp. 115–35. New York: Springer.

Ball, D.L., Hoyles, C., Jahnke, H.N. and Movshovitz-Hadar, N. (2002) The teaching of proof, in L. I. Tatsien (ed.) *Proceedings of the International Congress of Mathematicians*, Vol. III, pp. 907–20. Beijing: Higher Education.

Basinger, D. (1979) More on flow proofs in geometry, *The Mathematics Teacher*, 72(6): 434–6. https://www.jstor.org/stable/27961374

Bergwall, A. (2021a) Proof-related reasoning in upper secondary school: characteristics of Swedish and Finnish textbooks, *International Journal of Mathematical Education in Science and Technology*, 52(5): 731–51. https://doi.org/10.1080/0020739X.2019.1704085

Bergwall, A. (2021b) *Proof-related reasoning in upper secondary mathematics textbooks: Characteristics, comparisons, and conceptualizations.* Doctoral dissertation, Mälardalen University.

Fan, L., Mailizar, M., Alafaleq, M. and Wang, Y. (2018) A comparative study on the presentation of geometric proof in secondary mathematics textbooks in China, Indonesia, and Saudi Arabia, in G.Kaiser (ed.) *Research on Mathematics Textbooks and Teachers' Resources*, pp. 53–65. Cham, Switzerland: Springer.

Hadas, N., Hershkowitz, R. and Schwarz, B.B. (2000) The role of contradiction and uncertainty in promoting the need to prove in dynamic geometry environments, *Educational Studies in Mathematics*, 44(1): 127–50. https://doi.org/10.1023/A:1012781005718

Hanna, G. (2000) Proof, explanation and exploration: an overview, *Educational Studies in Mathematics*, 44(1): 5–23. https://doi.org/10.1023/A:1012737223465

Hemmi, K., Lepik, M. and Viholainen A. (2013) Analysing proof-related competences in Estonian, Finnish and Swedish mathematics curricula—towards a framework of developmental proof, *Journal of Curriculum Studies*, 45(3): 354–78. https://doi.org/1 0.1080/00220272.2012.754055

Herbst, P.G. (2002) Establishing a custom of proving in American school geometry: Evolution of the two-column proof in the early twentieth century, *Educational Studies in Mathematics*, 49(3): 283–312. https://doi.org/10.1023/A:1020264906740

Jones, K. (2000) Providing a foundation for deductive reasoning: students' interpretations when using dynamic geometry software and their evolving mathematical explanations, *Educational Studies in Mathematics*,44(1):55–85.https://doi.org/10.1023/A:1012789201736

Köğce, D., Aydın, M. and Yıldız, C. (2010) The views of high school students about proof and their levels of proof (the case of Trabzon), *Procedia-Social and Behavioral Sciences*, 2(2): 2544–9. https://doi.org/10.1016/j.sbspro.2010.03.370

Martinez, M.V. and Pedemonte, B. (2014) Relationship between inductive arithmetic argumentation and deductive algebraic proof, *Educational studies in mathematics*, 86(1): 125–49. https://doi.org/10.1007/s10649-013-9530-2

McMurray, R. (1978) Flow proofs in geometry, *The Mathematics Teacher*, 71(7): 592–5. Available at: https://www.jstor.org/stable/27961374 (accessed 6 September 2023).

Millar, R. (2011) Reviewing the National Curriculum for science: opportunities and challenges, *Curriculum Journal*, 22(2): 167–85. https://doi.org/10.1080/09585176.2011.574907

Miyakawa, T. (2012) Proof in geometry: A comparative analysis of French and Japanese textbooks. T.Y. Tso (Ed.), Proceedings of the 36th Conference of the International Group for the Psychology of Mathematics Education, Vol. 3, pp. 225–32. Taipei, Taiwan: PME.

Miyazaki, M. (2000) Levels of proof in lower secondary school mathematics, *Educational Studies in Mathematics*, 41: 47–68. http://dx.doi.org/10.1023/A:1003956532587

Oates, T. (2011) Could do better: Using international comparisons to refine the National Curriculum in England, *Curriculum Journal*, 22(2): 121–50. https://doi.org/10.1080/0 9585176.2011.578908

Pedemonte, B. (2007) How can the relationship between argumentation and proof be analysed?, *Educational Studies in Mathematics*, 66(1): 23–41. https://doi.org/10.1007/ s10649-006-9057-x

Pedemonte, B. (2008) Argumentation and algebraic proof, *ZDM Mathematics Education*, 40: 385–400. https://doi.org/10.1007/s11858-008-0085-0

Rind, A.A. and Mughal, S.H. (2020) An analysis of Pakistan's national curriculum of mathematics at secondary level, *Electronic Journal of Education, Social Economics and Technology*, 1(1): 39–42. https://doi.org/10.33122/ejeset.v1i1.4

Soto-Johnson, H. and Fuller, E. (2012) Assessing proofs via oral interviews, *Investigations in Mathematics Learning*, 4(3): 1–14. https://doi.org/10.1080/24727466.2012.11790313

Stylianides, A.J. (2007) Proof and proving in school mathematics, *Journal for research in Mathematics Education*, 38(3): 289–321. https://doi.org/10.2307/30034869

Stylianides, A.J. (2019) Secondary students' proof constructions in mathematics: the role of written versus oral mode of argument representation, *Review of Education*, 7(1): 156–82. https://doi.org/10.1002/rev3.3157

Toulmin, S.E. (1993) *The use of arguments, updated edition*. Cambridge, England: Cambridge University Press. French translation: De Brabanter P. *Les usages de l'argumentation*. Paris: Presses universitaire de France; DL 1993.

Winer, M.L. and Battista, M.T. (2022) Investigating students' proof reasoning: analyzing students' oral proof explanations and their written proofs in high school geometry, *International Electronic Journal of Mathematics Education*, 17(2): 1–21. https://doi.org/10.29333/iejme/11713

Part 2

Maths Lesson Planning Frameworks

3 Lesson planning and the mastery approach

Key arguments

Lesson planning in secondary mathematics takes shape as a natural outcome of teachers' knowledge *for*, *in*, and *of* practice. In this chapter, we argue that the focus of lesson planning should be combining and valuing interconnectedness with the web of mathematical knowledge. This can only achieved through reasoning. Therefore, we hold up reasoning as the overarching domain for lesson planning, where it involves pedagogical reasoning and teachers' reflections on their pedagogical reasoning.

Key terms

- Lesson planning

  Lesson planning is a valuable stage in the teaching and learning of mathematics, one which sets out what *ought to* work in the learning process. When the lesson planning focus is the external product, the planning may employ a goal-oriented approach, one that is driven by testing pressures. Or planning might harness a structure-oriented approach with the aim of unfolding mathematical concepts to develop deeper understanding. But if the focus of the lesson planning is the process of refining and reshaping pedagogy, then it can be understood as a process. Either way, pedagogical reasoning appears whenever there is lesson planning.

- Pedagogical reasoning

  This is about what should be learned, and how and why it should be done in this way. The mastery approach, for example, is conceptualized in terms of big ideas such as coherence, representation and structure, mathematical thinking, fluency, and variation. Each idea is a necessary part of the picture, and none exist without the others. The interaction of conceptual and procedural variation affect every part of the lesson.

The beginning of this chapter explores how lesson planning – understood as hypothetical teaching and learning trajectories in the mathematics classroom – is currently conceptualized from two perspectives: a product or a process. Then, the chapter zooms into the mastery approach in mathematics lesson planning, which is part of current UK practice. It also looks at the foundations of the mastery approach from variation theory (also known as *bianshi* 变式 theory) in China. Finally, the chapter discusses pedagogical reasoning from a theoretical point of view, proposing two practical ways to bring reasoning into lesson planning. This is about identifying the unifying principles that flow through a lesson, while being mindful of the differences in contextual factors that lead to richness in learning opportunities.

3.1 Lesson planning

When talking about lesson planning, what might emerge in the mind is a picture of a school's lesson plan template, something that asks for details on learning objectives, key mathematical literacy, SEN (special education needs) student support plans, cross-curriculum links, etc. Some may think of a school-style PowerPoint presentation, or a mental script begging to be filled up with interesting mathematical activities and questions to engage all students. Documents like this lay out a framework for how the lesson fosters the learning journey for all. Then, some teachers might wander into resource websites or textbooks for inspiration, hoping to find beautifully organized activities waiting for them.

Especially at very beginning of a teaching career, trainees commonly subscribe to the equation that lesson planning = lesson structure = resource hunting = differentiation practices = printing activities = time consuming. This is one simplistic way to summarize the action points involved in lesson planning. Trainees in England have been found to prioritize contextual factors, such as students' individual needs, as the most important element in lesson planning (John 1991). But this may fail to fully account for the hypothetical learning trajectory upon which both the teaching experience and the student learning experience will be based.

From the beginning of this chapter, we take a step back to look at how lesson planning is conceptualized, comparing the work of novice and experienced teachers across different countries. We summarize the three main approaches observed. We start from a goal-oriented approach, highlighting one particular challenge that exists across many education systems: testing, something which carries high stakes for both teachers and students. Then, turning to the structure-oriented approach offers a look at how concrete rules of one kind or another may shape planning. One theme we pursue particularly is understanding the differences between *what ought to* work in a lesson and *what actually works*, the differences between planning and delivery. The final dimension of the process-oriented approach we look at describes lesson planning influenced by the reflection that comes with peer observation, placing ever more emphasis on developing students' thinking and learning.

3.1.1 A goal-oriented approach

All lessons, whether in mathematics or other subjects, are situated in the classroom culturally and socially (Torff and Sternberg 2001). But equally, they are surrounded, to some degree or another, by pressures from the policy environment – such as the high-stakes testing that forms part of the education systems of different countries. In contexts of high-states testing, such as in the United States, how to best prepare students for mathematics assessments looms large in the lesson planning decision-making process – more so for experienced teachers than for novice teachers (Amador and Lamberg 2013). This means the fundamental focus for lesson planning becomes teachers' hypotheses regarding test preparation and their past experiences with student achievement in particular curriculum contents – rather than students' mathematics knowledge and the learning journey itself.

Academic success In fact, this is unavoidable in education systems with high-stakes testing; indeed, students' academic achievement is undeniably important. But research conducted in Turkish high schools (Mouratidis et al. 2018) compared this kind of performance-goal approach, with its focus on students outperforming others, with a mastery-goal approach – focusing on attaining mastery learning (we will discuss mastery approaches in the next section). In terms of learning outcomes, both approaches are directly associated with academic success. But while the mastery goal brings the additional value of challenge-seeking, high-stakes testing permeates the instructional content through a didactic style often being forged (Diamond 2007). Nevertheless, sometimes lessons transcend the development of mathematical understanding and devote themselves to the practice of reflective teaching.

3.1.2 A structure-oriented approach

Lesson planning mainly requires a strong sense of structural perspective when unfolding mathematics concepts and stimulating the development of understanding. To investigate this, we start with the use of curriculum materials, including textbooks, which might provide the highest level of structure in lesson planning. This then moves towards a lesson structure based on problem-solving, with its strong sense of step-by-step development, and finally a non-linear structure, which features a loose analytical style.

Curriculum materials-based structure Use of mathematics textbooks to structure lessons began decades ago around the world. The important role of textbooks in teaching and learning mathematics has been identified as offering a step between the national curriculum and the classroom (Fan, Zhu and Miao 2013). The verb best capturing this spirit is *to use*, as in *using* the textbook, rather than *teaching* the textbook (Fujii 2018; Matić 2019). Using a textbook during the lesson planning stage means the 'classification of teaching purposes', or setting learning objectives, and 'structuring and sequencing' the main activities to arrive at an overall structure (Fan et al. 2021, p. 1321). Critics of these uses of textbooks, however, have also been around for decades. They question whether these textbook uses impact adversely upon the expression of teacher identity and minimize teacher engagement with other resources

(Collopy 2003), especially for beginner teachers (Ball and Feiman-Nemser 1988).

With the rapid development of classroom technologies, ready-made sets of curriculum materials, including lesson slides, have become widely available in the UK. Those such as the Mathematics Mastery Programme (www.arkcurriculum.org.uk) or White Rose resources (https://whiterosemaths.com/resources) are supposed to serve similar purposes to paper textbooks. The Maths Hub lead Nicola Fareham (N. Fareham 2022, personal communication, January 27) expressed concerns about use of these resources for mathematics teacher trainees, suggesting that it might undermine the lesson planning process. On the other hand, the argument Tummons puts forward in support of textbook usage could also apply here: they are 'a focus for inquiry' (Tummons 2014, p. 421), bringing a structured approach to lesson planning. At the centre of such a structured approach is the aim of mastering the content, with all aspects of the teaching steered through the lesson planning towards the ultimate objectives set by the curriculum materials.

Step-by-step structure Problem-solving lesson structures from Japan have been widely implemented around the world, including in Australia (Groves 2013), South Africa (Chirinda and Barmby 2018) and the States (Lewis 2002). When comparing how pre-service and in-service teachers plan the problem-solving structure of their lessons in Japan, the main commonality found is a step-based conception, along with anticipation of the possible solutions students will come up with (Shimizu 2008).

Regardless of different cultural backgrounds, curriculum realities and education systems, this problem-solving style of lesson structure tends to start from a single problem, one connected to lesson objectives and students' mathematical development. A typical problem for multiplication lessons at lower-primary level runs like this: 'I bought five chocolate bars, and each one has four pieces. How many pieces do we have? Do we have enough for everyone to have a piece?' (Lewis 2016, p. 10). Then students work either individually or collaboratively on the solutions, while teachers guide them to compare and refine their answers. At the end of the lesson, teachers sum up what has been learned, with the intention of promoting a higher level of mathematics thinking.

There are actually two elements worth noting within this approach to lesson planning: (1) the key questions to ask, making prediction of how students may respond (beyond just the desired solutions), and teachers' self-evaluations (Doig and Groves 2011); and (2) the problem itself to be used in the lesson. To prepare for this lesson, one part to consider is how to choose the problem and how to pose it with the classroom culture in mind, considering students' motivations and the nature of the mathematics task. This has led to significant academic research in the area of task design (Watson, Ohtani, and Ainley 2015). The underlying belief is in the necessity to maximize student engagement

(Pang 2016), seeking a rigorous and meaningful way to promote problem-solving and communication.

With this approach, students' learning experience becomes a much more important consideration in the lesson planning. And because of the ever-changing nature of the classroom, an approach relying on fixed content-centred curriculum material is not always appropriate. Rather, it is possible to rely on the flow from step to step when seeking to establish principles and to generate rules for how teaching and learning happen, alongside predictions about what kind of learning should come next to best engage students.

Non-linear-based structure This approach involves several interconnected components of lesson planning – far from the focus on fixed content from curriculum materials, or geared towards the development of problem-solving skills. Based on the theory of constructivism and emphasizing learner experiences, Lowrie and Patahuddin (2015) summarized a structure meant to guide this non-linear process of lesson planning: experience-language-pictorial-symbolic-applications (ELPSA). *Language*, including communication in the classroom, promotes sense-making around mathematical ideas. Concrete and visual representations facilitate progression towards the understanding of the *pictorial*, eventually moving towards abstract *symbolic* representation for deep conceptual understanding. Finally, students are able to *apply* what they know to a new situation, including those based on real-life episodes. The framework has a recursive-progressive nature (Lowrie, Logan and Patahuddin 2018). Teaching for mastery, which we will discuss later in the chapter, draws on similar underlying principles.

Now, if we take the learner's perspective, explicitly with the learner's experience in mind, research has detected patterns in the active learning processes of *noticing*, and also *productive* noticing. In his thesis, Choy (2015) proposed the FOCUS framework, which highlights productive noticing at two levels:

1 The FOCUS, as a noun: an explicit focus on *key points* of the mathematics concept, *difficult points* for highlighting confusions, and *critical points* in the teacher's course of action, such as an instructional decisions.
2 To FOCUS, as a verb: designing lessons that reveal students' thinking and promote mathematical reasoning.

Here, predicting student thinking becomes the basis of the lesson planning – even though novice teachers might not find this easy (Ivars et al. 2018). Teachers are supposed to review the lesson they have taught and justify their instructional decisions through pedagogical reasoning (Choy, Thomas and Yoon 2017). Therefore, lesson planning is associated with reviewing a set of experiences for teachers, for students, between teachers and students, and between students and students – rather than being purely a matter of directly pursuing knowledge development for learners.

3.1.3 A process-oriented approach

This approach shifts lesson planning from focusing on an external product to a process of collaborative professional development for teachers. This means considering lesson planning as a practice, as part of a dialogical model, rather than merely as an end product, one represented by actions such as filling in a lesson-plan template (John 2006). In line with this view, mathematics education research has paid growing attention to Japanese Lesson Study (Isoda 2007) and Chinese Exemplary Lesson Development (Huang, Su and Xu, 2014; Huang and Bao, 2006). Both illustrate particular ways to establish the source of a good lesson (and good lesson planning), and how to tailor the teaching to local students' learning needs. Both rest on collaboration within the teaching community and draw on the experience and expertise of other teachers, embedding these in classroom teaching and learning realities.

Lesson Study Cycle in Japan It consists of (1) identifying the focus; (2) planning in collaborative groups; (3) conducting a research lesson; (4) holding post-lesson discussion within the group; (5) repeating cycles of research with new lessons; (6) employing outside expertise, and (7) mobilizing knowledge (Seleznyov 2018). Tall (2008, p. 48) noted that such lessons involve 'a steady compression of knowledge' with 'explicit attention to the essential ideas'. High-quality lesson planning sets up progress by identifying key milestones with sequences of tasks, and making interconnections with previous tasks (Mendes, Brocardo and Oliveira 2021). It provides teachers space and opportunities for reflection on their practice (Lewis 2016).

Exemplary lesson development cycle in China It is known as *keli* (课例) in China, displays a very similar pattern to a Japanese lesson study cycle. Lesson planning here follows three phases (Huang and Bao 2006):

1 Familiarization and focusing to decide upon the target of the lesson and upon relevant content.
2 A cycle of teaching, reflection and revision, which includes three teaching stages (existing action, new design and new action) with two stages of reflection in between meant for updating ideas and improving action.
3 Disseminating the *keli* process and exemplary lessons.

Regarding the topic of Pythagoras' theorem, for example, through expert contributions in the *keli* group, lesson planning might add scaffolding for discovery and proof of the theorem, concentrating on how to find the proof rather focusing on the proof itself, $a^2 + b^2 = c^2$. Within this process, the various reflections and redesigns of previous lesson(s) with the same learning objectives allows participating teachers to feed different contributions into a teaching approach that fits his or her classroom (Huang et al. 2019). These are built on collaboration between participants with diverse backgrounds, and the space for being reflective and reflexive.

Approach The general consensus is that lesson study or exemplary lessons improve awareness of how to teach while building the ability to reflect. But precisely how has this kind of lesson planning developed over the years? Huang and Li (2009) recognized four different approaches emerging in the development of Exemplary Lessons.

It started out with a *daily situation-oriented design*, which emphasized the context of daily life to engage students – arguably at the expense of mathematical essence. A second stage, *developing a concept-oriented mathematics design*, paid great attention to 'forming, developing and consolidating' the concepts being studied, with the possible price of losing student interest. The third, *a balanced design of mathematical concept formation and student participation*, successfully achieved both mathematical understanding and student engagement, but it lacked particular care for effective time management. The final design, *multiple effective teaching methods, not a fixed teaching method*, optimized opportunities to achieve learning objectives in the Chinese context.

Taken together, these designs are examples of how teachers can access *knowing* through *being*, shifting the emphasis from what teachers know to what they become, to a process (Scheiner et al. 2019). While this has worked well in the Chinese context, things might not be so straightforward in the UK, as underlying these approaches is the recognition of the importance of lesson planning. In the case of a reasoning project the authors carried out in Durham in 2020, we found a particular challenge lies in carving out time and space for continued professional development that enables teachers to have an experience of how important the lesson planning is, and to make it part of the impulses that lead to their choices and actions (Wang and Brown 2021). Before looking at the mastery approach, which is seen as a key factor in the successes of mathematics education in Shanghai, we first summarize key points about lesson planning in box 3.1, as a synthesis of the discussion so far.

Box 3.1 Lesson planning

- is a (collaborative) process in which teachers reflect on their practice
- employs a structural perspective to unfold mathematics knowledge via the active process of pedagogical reasoning
- presents tasks in a logical and interconnected way
- needs space to maximize students' engagement.

3.2 The mastery approach to lesson planning

The term 'mastery' has come to summarize the teaching approaches that are seen as a key factor in the successes of mathematics education in Shanghai, and the pedagogical theory behind it is now being adopted and adapted for use in the UK's classrooms. Indeed, since 2014, applying 'mastery' methods has become a policy solution for raising performance standards (or, at least, PISA league table positions). In 2009, Shanghai took part in the PISA exercise for the first time, and its students' top performance in maths, science and reading was considered a stunning success. It demonstrated the students' abilities in conceptualizing, generalizing and creatively using information based on their own investigations and modelling (OECD 2010).

Since student learning outcomes are among the most highly valued ends, some schools in England have tried changing their lesson-plan templates to fall into line with lessons from Shanghai. The main body of the template now includes four columns: (1) time, (2) learning activity, (3) assessment opportunities/differentiation/questioning techniques, and (4) success criteria/student outcomes. A goal-oriented approach is thus manifest in efforts to measure the learning outcomes of each lesson, and in generating verdicts about whether students have improved over time.

In this section, we will take a historical perspective on the term 'mastery' and outline the predominant framing of mastery approaches in the UK. Then, we will explore its roots in pedagogical theory emerging from Shanghai, variation or *bianshi* theory. This knowledge is intended to enable recognition of the pedagogical reasoning that drives the mastery approach, and to help in identifying areas useful in evaluating one's own lesson planning.

3.2.1 Teaching for mastery

For most of the past half-century, educators accepted Bloom and his colleagues' taxonomy framework (Bloom 1956), which categorized six educational goals: knowledge, comprehension, application, synthesis and evaluation. More recently, this framework has been refined to help navigate six levels of cognitive process: remembering, understanding, applying, analysing, evaluating and creating (Krathwohl 2002). But besides this well-known taxonomy, Bloom (1968) also proposed the idea of learning for mastery, which consisted of:

1 A belief that most students, for example, over 90 per cent of students, can master the subject.
2 The assessment strategy/system to achieve this level of subject mastery, e.g. parallel and regular formatives.
3 Pedagogy based on group instruction.
4 Consideration of how individuals may respond to this pedagogy, including the time needed for a certain topic and the pace of progress.

This by no means implied something top-down and imposed by teachers, but rather it was a self-constructive approach for learners. These ideas have been taken up across subjects by those willing to believe that a mastery level is achievable for the majority.

The term mastery returned to our attention again, at the level of UK education policy, when the England-Shanghai Teacher Exchange first took place in 2014. The exchange involved 71 primary teachers from England and 59 primary teachers from Shanghai. The NCETM (National Centre for Excellence in the Teaching of Mathematics) hosted the Exchange Project face to face up until 2019 (when the COVID-19 pandemic period began). In these exchanges, mathematics teachers on both sides have enjoyed opportunities for teaching, observation, and subsequent in-depth conversation (Yuan and Huang 2019). The primary intention has been to make sense of teaching and learning mathematics at the classroom level on the other side of the world, and to explore what ideas can be transferred, to what extent and in which ways. The biggest complication in making sense of the mastery approach, though, are factors from outside the classroom influencing teaching and learning.

Let's distinguish three major factors: textbooks, culture and issues in the global transmission of education policy. First, recognition of the quality of textbooks in Shanghai led to the announcement of UK government funding for textbooks employing a mastery approach (DfE 2016). But cultural differences are manifest both in ideological perspectives, including beliefs, and structural aspects, such as the organization of mathematics teaching (Boylan et al. 2019). Elliott (2014) raised concerns about the transition of pedagogy from one context to another, since the supporting elements found in one culture and context might be missing in another. And Clapham and Vickers' (2018) offered insights into how to implement policy interventions under these circumstances. But with all of these caveats in mind, let's take a look at how this particular lesson planning approach, especially its structure orientation, has unleashed the potential of mastery in mathematics.

A structure based on mastery curriculum materials The Ark multi-academy trust has 39 schools in Birmingham, Hastings, London and Portsmouth. Following the visit of a group of teachers from Ark schools to Singapore in 2012, they developed a new mathematics curriculum, the Mathematics Mastery Programme. Later, it has become a commercial package, featuring schemes of work, planning guidance, lesson resources, and assessment and intervention tools, as well as integrated professional development (Mathematics Mastery, n.d.). These materials go right down to the level of everyday classroom exercises, with the mastery approach defined by tasks and activities shown in these slides, and intended to be tailored by schoolteachers to fit with their particular group of students. While the Mathematics Mastery Programme covers primary and secondary settings, two textbooks developed with the help of DfE funding are particularly popular in primary schools: *Maths-No Problem!*, and *Power Maths*. This brings us to another side of the mastery approach – using a programme/textbooks titled or rubber stamped by the government. This has demonstrated the meaning of mastery approach in daily practice.

A step-by-step mastery structure   A concrete-pictorial-abstract (CPA) approach has been seen as a key component of the Singapore Mathematics Curriculum, believed to be behind the levelling up of students' understanding to the abstract level (Naroth and Luneta 2015). Underlying the theory of the CPA approach is Bruner's notion of the structure of knowledge, which involves three stages: enactive, iconic and symbolic (Bruner 1966). This approach has been widely used to deepen understanding of topics such as fractions (Africa et al. 2020) and circumferences (Salimi et al. 2020). It lays down the broad sequence for the instructional structure, one that is not only effective but also adaptable for use in different contexts (Leong, Ho and Cheng 2015).

Researchers in STEM subjects have also found learning sequences, such as the Systematic and Integrative Sequence Approach (SISA), to be effective in supporting learning around complex concepts (Tíjaro-Rojas et al. 2020). This approach is based on Bloom's Revised Taxonomy, featuring six sequential steps, and it highlights a crucial principle of this inductive approach: the systematization and integration of related concepts.

A non-linear-based mastery structure   Although there is no single clear conception of mastery in mathematics (Simpson and Wang 2022), the predominant explanation is offered by the NCETM. It is known as the 'five big ideas in teaching for mastery' (see Figure 3.1): coherence, representation and structure, mathematical thinking, fluency and variation.

Figure 3.1 Teaching for Mastery

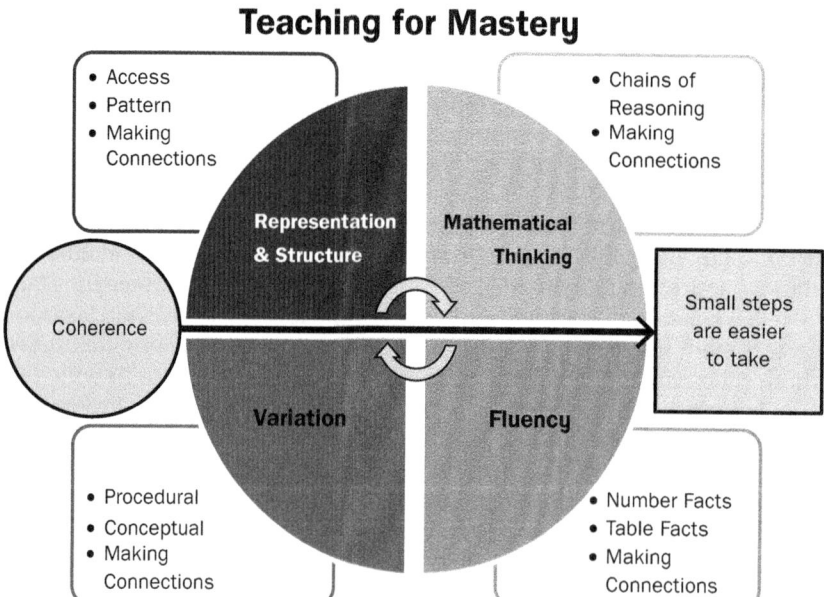

Teaching for Mastery

- Access
- Pattern
- Making Connections

- Chains of Reasoning
- Making Connections

Representation & Structure   **Mathematical Thinking**

Coherence

Small steps are easier to take

Variation   **Fluency**

- Procedural
- Conceptual
- Making Connections

- Number Facts
- Table Facts
- Making Connections

(NCETM, 2017)

Here, coherence means providing small steps as scaffolding to master a concept, ranging from those aiding understanding of the concept's meaning to those concerned with its application in a range of contexts. Representations of concrete, pictorial and abstract kinds are used to unfold the mathematical structure within the concept itself as well as links with other concepts. If the first two ideas, coherence and representation and structure, shape teachers' lesson planning, then the next two, mathematical thinking and fluency, emphasize the importance of learners' cognitive development in mathematics. Students showing success in these two elements are able to systematically rationalize the relevant mathematical concepts and procedures within different contexts via reasoning with others. The last idea, variation, occupies a central position in lesson planning: a recognition of the indivisible aspects of a certain concept, and their interdependence with activities in the development of understanding.

3.2.2 Variation theory/*bianshi* 变式 teaching

The idea of *bianshi* teaching through variation was proposed by Professor Gu Lingyuan in the 1980s, as a summary of an effective teaching method to promote deeper understanding of mathematics in the context of mixed-ability students and large class sizes in Shanghai (Gu, Huang and Gu 2017). The outstanding performance of Shanghai students in PISA from 2009 has been seen as the result of this effective pedagogy, at least from a UK perspective. There are two core elements, conceptual variation and procedural variation. The purpose of conceptual variation is to establish a web of knowledge and to value interconnectedness among mathematics concepts.

Conceptual Variation Students' conceptual understanding is developed through recognizing the uniqueness of the particular concept being studied – its critical features – and also its links with others. The underlying belief is a mechanical view of mathematics, with concepts able to be split into different sub-concepts or properties. For example, to distinguish the geometric figures in 2-D shapes, students are asked to precisely determine the differences between these quadrilaterals (see Figures 3.2 and 3.3).

This not only highlights the structure of the concept but also the different types of representation possible. For example, the concept of perpendicular height in a triangle can be expressed pictorially in different types of triangle, as well as in non-standard figures of triangles. The non-standard or non-prototypical figure can bring new opportunities to the discussion, highlighting critical features, sub-concepts or properties that gradually become context-free.

Figure 3.2 Naming quadrilaterals

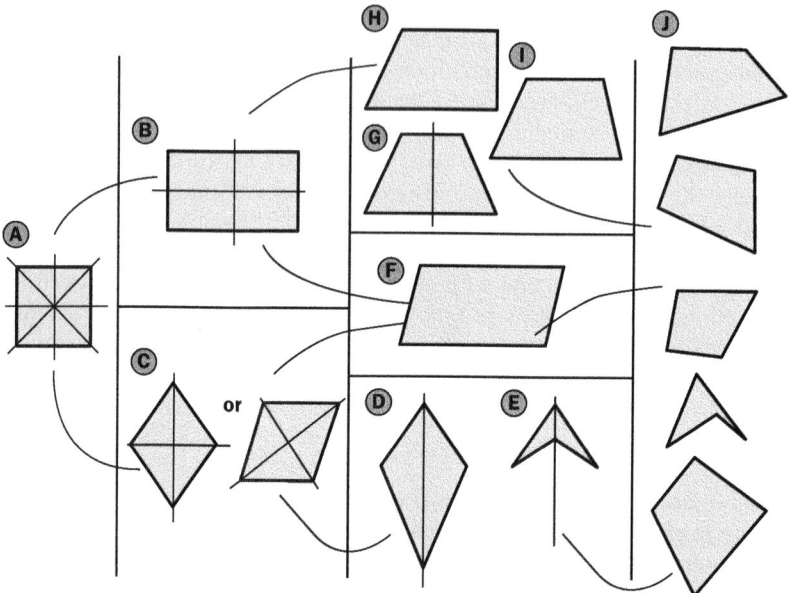

Figure 3.3 Classification of quadrilaterals – properties of each quadrilateral type

	Shape	Lines of Symmetry				All sides equal	All angles equal	2 sets of parallel sides	2 pairs of equal angles	2 pairs of equal sides	1 set of parallel sides	1 pair of equal angles	1 pair of equal sides	Order of Rotational Symmetry			Diagonals	
		4	2	1	0									4	2	1	Equal	At 90°
A	Square																	
B	Rectangle																	
C	Rhombus (Diamond)																	
D	Kite																	
E	Delta																	
F	Parallelogram																	
G	Isosceles Trapezium																	
H	Right Angled Trapezium																	
I	Trapezium																	
J	Irregular Quadrilateral																	

What additional properties do all quadrilaterals have which have not been stated in the table?
(i.e. what makes a quadrilateral a quadrilateral)
Do you know of any other 'type' of quadrilateral which hasn't been talked about - if so what special properties does it have?

Procedural variation There are three major functions behind procedural variation: forming a concept, solving problems and establishing a specific experience
system through a series of variations (Zhang, Wang and Kimmins 2017). Procedural variation takes an analytical approach for looking at how a task, or a series

of tasks, could be designed to accumulate deeper understanding. There are different kinds of task possible here: variation on a problem, one question multiple solutions (such as the different methods of multiplication) and multiple questions one solution (Sun 2011). The NCETM calls these sequences of tasks 'intelligent practice', since they are intended to foster fluency without repeated practice (Blausten et al. 2020). Procedural variation follows a scientific method, progressively unfolding mathematical activities. As such, the purpose of procedural variation is to build increasing levels of challenge, including when a task calls on different methods.

Variation lesson planning may be expected to intertwine these two kinds of variation with a structure-oriented approach. A typical variation lesson could be structured as follows (Huang and Leung 2017):

1   Reviewing and introducing by using procedural variation 1 (to start the process of procedural variation) – reviewing previous knowledge and bringing the new topic to students' awareness.
2   Exploring new concepts, with conceptual variation 1 employed for describing new concepts; conceptual variation 2 for different orientations of the basic/standard/prototypical figure.
3   Examples and exercises, where procedural variation 2 takes place for different contexts of prototypical figures, and procedural variation 3 is put towards different ways of applying new concepts; conceptual variation 3 for contrast and counter-examples.
4   Summary and assignment, with procedural variation 4 used for creating opportunities for learning a new topic; conceptual variation 4 for consolidation and memorization of the concepts.

Reflecting on the opening discussion of chapter 1, featuring Plato and Aristotle, in this example, conceptual and procedural variation could also be viewed as a growth phase, one featuring reflection on the objective structure of these concepts and their position in the web of mathematics knowledge and in subjective experiences such as problem-solving. In the course of this growth, as variations feed into mathematical knowledge and the associated learning journey, things can feel chaotic; cultural factors, systemic factors, important school factors, and key classroom factors can have significant impacts on the application of variation theory.

It is also worthwhile to note that Ference Marton and his colleagues have proposed a variation theory of learning (Marton 2015), a general theory originating from phenomenography traditions rather than from maths-specific ways of thinking (Kullberg, Runesson Kempe and Marton 2017). This variation theory of learning focuses on how to develop learners' capabilities to discern the key aspects of a situation, especially a novel situation, through variation. There are four patterns: contrast, separation, fusion, and generalization. These four patterns arise in the exploration of the differences (variance) and similarities (invariance) in different phenomena.

The question, then, is what exactly do the *patterns* in this variation look like – and how does the opposite, *invariance*, come into it? Peng, Bao and Ki (2017) summarized two main patterns that enrich the breadth and depth of understanding: separation and fusion. These patterns go beyond understanding of a particular piece of knowledge and its applications, and beyond identifying sameness (invariant structure) and differences (on the basis of invariant structure).

There are many causes to believe that reasoning is involved in these two main patterns. For example, backwards and forwards reasoning around sameness and differences is a key part of generalizing around invariance. However, there is little discussion on the role of reasoning within this variation theory, especially when it comes to procedural variation. In the next section, we will look at pedagogical reasoning, which underlies how lesson planning is formed. We will consider why this particular way might be employed to organize a hypothetical learning trajectory, in the form of a decision-making process shaped by clear intentions. We shift the focus from what the lesson planning looks like and how to structure it, towards questions of 'why'.

3.3 Pedagogical reasoning

3.3.1 What is pedagogical reasoning?

While student feedback gives us some control over the quality of teaching, pedagogical reasoning shines a brighter light upon the unseen thinking process occurring with students' minds. We have already discussed a process-oriented approach to lesson planning, offering the examples of Japanese Lesson Study and Chinese Exemplary Lesson Development. The importance of process found here is about making the values and beliefs held by the teacher about the mathematics topic itself, and about the learning context, explicit in consciousness. Loughran (2019, p. 526) proposed that 'teachers' knowledge *for, in* and *of* practice' can be viewed as the foundation of professional knowledge in teaching. Here, knowledge *for* practice is a form of theory, such as that about self-efficacy – theory with a big T. Knowledge *in* practice refers to teachers' reflections on their experiences with theory – theory with a small T. Knowledge *of* practice is akin to practice but with formalized processes for generating knowledge – for example that gained through action research or small-scale research projects. Pedagogical reasoning brings these three types of knowledge together to show what teachers know, how they know what they know, and why certain things work in practice.

It is important to be clear that lesson planning sits in a broad pedagogical-reasoning framework. The underlying general educational questions of this framework include: what should be learned, how, and why should it be done in this way? And what constitutes an educated person (Webb 2002)? Shulman (1987) expressed awareness that teaching begins with reasoning, both as a noun and verb. It starts from comprehension of a set of ideas to be taught (with the consideration of related ideas within the subject, with ideas in other subjects, knowledge of educational purposes etc.), while reasoning follows as a transformation of the knowledge (including knowledge about learners) and reflection on ideas, values and beliefs about knowledge and about learners. The model Shulman proposed features six steps: comprehension, transformation, instruction, evaluation, reflection, and new comprehensions (Shulman 1987, p. 15). The steps are about reaching moments of 'pedagogical equilibrium' (Mansfield and Loughran 2018), a similar idea to the 'cognitive equilibrium' (Piaget 1952; 1977; 1985) proposed by Piaget. These two ideas are about the tipping moment which justifies the teaching approach, a moment of shifting from *telling* to *teaching*, and where one dares to test if an alternative teaching approach works, and to what extent.

3.3.2 Pedagogical reasoning in practice

Route 1: Balancing conceptual understanding through maths in real-life (Inquiry Maths) and maths itself The UK classroom normally creates an environment of support and inquiry. Such a strength could be made even more effective for student outcomes by also fostering an emphasis on the mastery approach. This means *teaching with purpose*, through a particular concern for achieving deeper understanding of the concept at hand, developing logical thinking, and strong applications. Here, memorization, and didactic teaching play important roles.

In the example of teaching gradients in line graphs, we normally refer to a visual impression of how steep the straight line is, and we use the analogy of 'along the corridor, up the stairs'. The power of the visual approach as part of the thinking process, and of graphic representation as a product of this process, is deeply ingrained in the normal teaching process. But this can be problematic when dealing with non-homogeneous systems (see Figure 3.4). In this example, the gradients of the two lines have the same value, even though the lines do not appear to have the same degree of steepness.

Figure 3.4 The gradient of the same function $y = 2x$ in two visual approaches

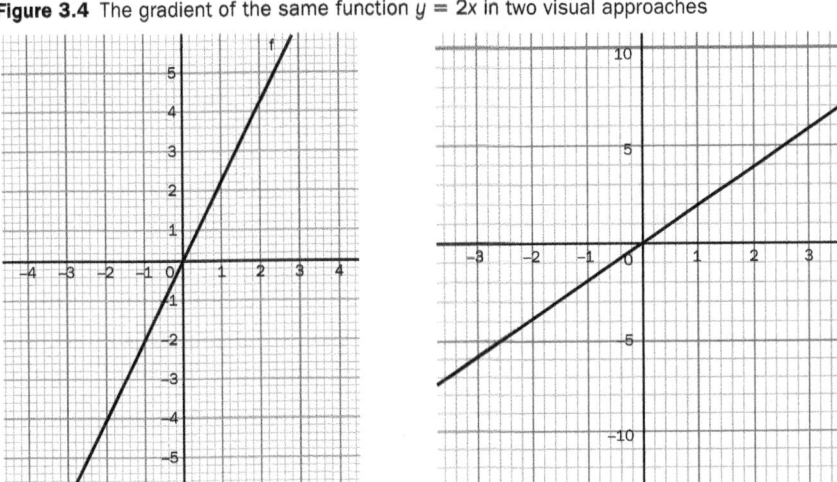

The way in which gradient is explained needs to be clarified using an analytical approach; an algebraic method of solving simultaneous equations (see Figure 3.5) brings precision through a numerical answer.

Figure 3.5 An algebraic approach to gradient

Put the two pair points Into algebraic expression: $y = ax + b$, $(a \neq 0)$.

$$\begin{cases} y_1 = ax_1 + b \quad (1) \\ y_2 = ax_2 + b \quad (2) \end{cases}$$

$(1) - (2)$: $y_1 - y_2 = ax_1 - ax_2$

$$y_1 - y_2 = a(x_1 - x_2)$$

$$a = \frac{y_1 - y_2}{x_1 - x_2}$$

It is crucial that teaching recognizes the challenges arising when students' intuitions conflict with definitions being taught; concepts are not always physically accessible to students in their maths learning. After one or two years of secondary school learning, consideration should be paid to how students develop mastery at abstract levels, promoting their awareness of the understanding to be gained from manipulating algebraic expression.

Route 2: Balancing student-centred and content-centred approaches Another strength of the UK classroom involves grasping what students can and cannot do, evaluations of the reality of what is possible to achieve and how to accommodate differing abilities. On the other hand, we need to shift some priority towards mastering specific concepts and their particular learning difficulties. This balance requires teachers to act both as facilitator/explainer and instructor/explainer. Some schools have proposed a lesson structure with the following steps: starter – modelling 1– applying 1– modelling 2– applying 2– plenary. This is useful in helping teachers and trainees recognize the affordances of a lesson and the constraints of tasks. And based on this lesson structure, a holistic approach can emerge in teachers' selection and implementation of tasks.

Taking the example of a Year 9 first lesson on straight line graphs, teachers at a state school in County Durham demonstrate an effective way of balancing the two approaches, as emerged in the PANDA project.

Starter
What do you know about linear graphs? You have one minute to write everything you already know about linear graphs. You may use bullet points, a spider diagram, other diagrams. etc.

<u>Modelling – convert algebraic expression into graph form</u>
Plot a linear graph $y = 3x - 5$ using table and graph (see Diagram 3.1), then use
the exact same x values to plot $2y + 5x - 4 = 0$ (see Diagram 3.2).

Plotting linear graphs

Diagram 3.1

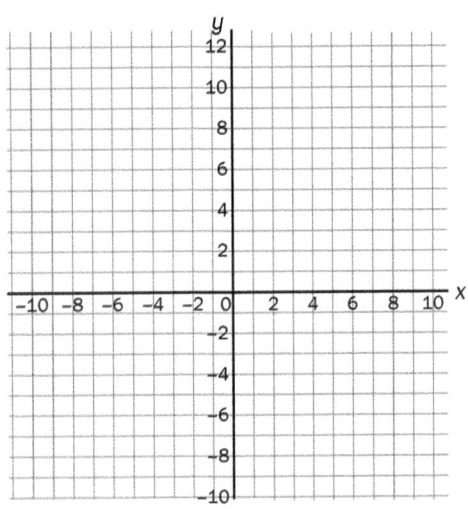

Plotting linear graphs

Plot: $y = 3x - 5$

x	-2	-1	0	1	2	3
y						

Diagram 3.2

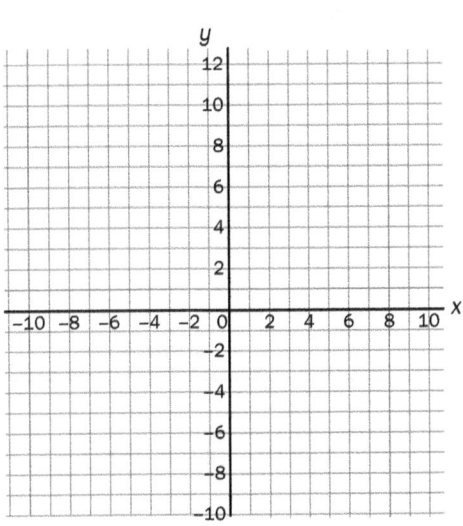

Plotting linear graphs

Plot: $2y + 5x - 4 = 0$

x	-2	-1	0	1	2	3
y						

Modelling – convert graph form to algebraic expression (see Diagrams 3.3 and 3.4):

Diagram 3.3

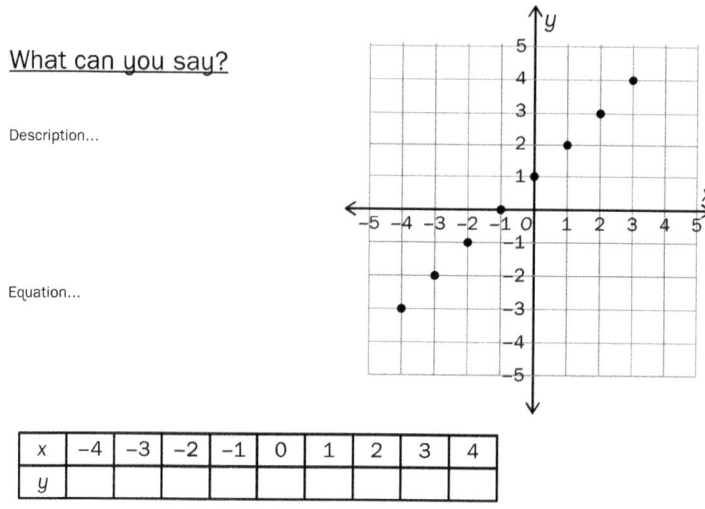

x	−4	−3	−2	−1	0	1	2	3	4
y									

Diagram 3.4

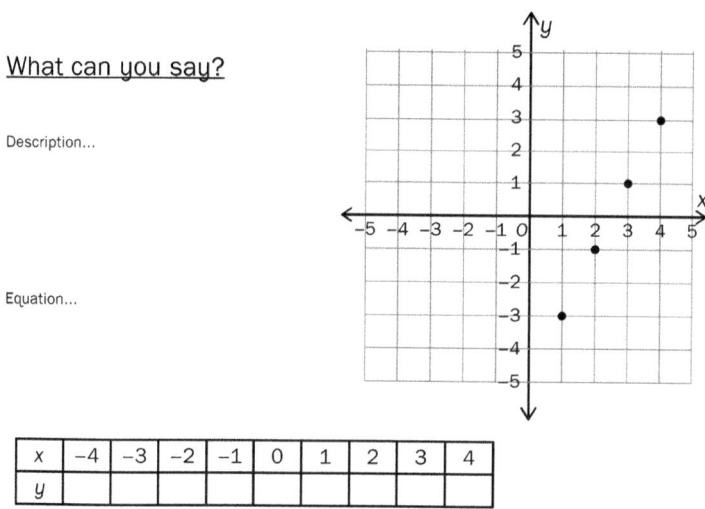

x	−4	−3	−2	−1	0	1	2	3	4
y									

Practice: Matching cards

Graph, equation of line, table of values, specific coordinate on a line, sequence, Nth Term

Enhancement

What happens if…

- The line $y = x + 2$ is reflected in the y axis?
- The line $y = 2x - 1$ is reflected in the x axis?
- The line $y = x$ is rotated 90° about the coordinates (1, 1)?

This lesson has clear intended goals, and suitable tasks were selected with procedural variation in modelling parts to make conjectures, and to develop and evaluate arguments. Tasks are linked with questions in the same format, with just one aspect changed: the different types of equations. This helps students to develop knowledge of the mathematical structure and to describe patterns, and it helps teachers focus on interaction with and among students, rather than only watching for what students can do. This is a very practical approach whose main focus is developing the opportunity for deep understanding. Furthermore, the practice and enhancement parts of the lesson consider both mathematical and pedagogical aspects of the tasks in connection with students' inquiry.

3.4 What comes next?

In this chapter, we have discussed three approaches to lesson planning: goal-oriented, structure-oriented and process-oriented. In particular, we have analysed mastery approaches to lesson planning from a UK perspective, as well as from that of Shanghai, China. Lesson planning can be thought of as like an iceberg, most of which is hidden deeply below the surface. The *product* of lesson planning is the visible tip of the iceberg, but this is the result of the *process* of lesson planning, which is based on pedagogical reasoning. It is not easy to measure or see, but more like an inner conversation.

Once we truly honour the idea that reasoning is not taught but facilitated – and facilitated through lesson planning – this puts into perspective that everyone has the ability to reason. When we promote reasoning, we are not *teaching* reasoning or *telling* reasoning. Instead, we should foster reasoning behaviour, through planning lessons that enable and encourage students, both explicitly and implicitly, to reason.

Developing thoroughly reasoning-focused lesson planning could help meet the challenges laid out in the introduction chapter. But to do so, we need a clearer understanding of how reasoning-focused lesson planning looks in practice, and how teachers experience this in their daily classroom. A possible way to combine both is the subject of the next chapter, the Causal Connectivity Framework.

Box 3.2 Top tips on lesson planning

- Lesson planning may tackle conceptual/procedural understanding and problem-solving, but reasoning should be the overarching domain.
- Lesson planning should focus on reasoning, via reflections and pedagogical reasoning.

Practice activities

Look at your recent lesson planning. Reflect on the types of lesson planning you have been using. List three examples of your own pedagogical reasoning; i.e. why it should be done in this way.

1

2

3

References

Africa, M., Borboran, A.M., Guilleno, M.A. et al. (2020) A lesson study on using the Con.crete-Pictorial-Abstract (CPA): approach in addressing misconceptions in learning fractions, *SABTON: Multidisciplinary Research Journal*, 1(1): 1–8. Available at: https://www.researchgate.net/profile/Arianne-Mae-Borboran/publication/356810071_A_Lesson_Study_on_Using_the_Concrete-Pictorial-Abstract_CPA_Approach_in_Addressing_Misconceptions_in_Learning_Fractions/links/61ae15c229948f41dbcde3fe/A-Lesson-Study-on-Using-the-Concrete-Pictorial-Abstract-CPA-Approach-in-Addressing-Misconceptions-in-Learning-Fractions.pdf (accessed 6 September 2023)

Amador, J. and Lamberg, T. (2013) Learning trajectories, lesson planning, affordances, and constraints in the design and enactment of mathematics teaching, *Mathematical Thinking and Learning*, 15(2): 146–70. https://doi.org/10.1080/10986065.2013.770719

Ball, D.L. and Feiman-Nemser, S. (1988) Using textbooks and teachers' guides: A dilemma for beginning teachers and teacher educators, *Curriculum Inquiry*, 18(4): 401–23. https://doi.org/10.1080/03626784.1988.11076050

Blausten, H., Gyngell, C., Aichmayr, H. and Spengler, N. (2020) Supporting Mathematics Teaching for Mastery in England, in F. Reimers (ed.) *Empowering Teachers to Build a Better World*. SpringerBriefs in Education. Singapore: Springer. https://doi.org/10.1007/978-981-15-2137-9_2

Bloom, B.S. (1956) *Taxonomy of educational objectives, handbook 1: The cognitive domain*. New York: David McKay Co Inc.

Bloom, B.S. (1968) *Learning for mastery: Instruction and curriculum*, Regional Education Laboratory for the Carolinas and Virginia, Topical Papers and Reprints, Number 1, Evaluation comment, 1(2). Available at: https://files.eric.ed.gov/fulltext/ED053419.pdf (accessed 20 May 2023).

Boylan, M., Wolstenholme, C., Demack, S. et al. (2019) *Longuitudinal evaluation of the Mathematics Teacher Exchange: China-England final report*, Department of Education. Available at: https://www.gov.uk/government/publications/evaluation-of-the-maths-teacher-exchange-china-and-england (accessed 20 May 2023).

Bruner, J.S. (1966) *Toward a theory of instruction* Vol. 59. Boston: Harvard University Press.

Chirinda, B. and Barmby, P. (2018) South African grade 9 mathematics teachers' views on the teaching of problem solving, *African Journal of Research in Mathematics, Science and Technology Education*, 22(1): 114–24. https://doi.org/10.1080/18117295.2018.1438231

Choy, B.H. (2015) *The FOCUS framework: Snapshots of mathematics teacher noticing* Doctoral dissertation, University of Auckland, New Zealand.

Choy, B.H., Thomas, M.O. and Yoon, C. (2017) Snapshots of productive noticing: Orchestrating learning experiences using typical problems. *Proceedings of the 40th Annual Conference of the Mathematics Education Research Group of Australasia*, pp. 445–66. Available at: https://repository.nie.edu.sg/bitstream/10497/22271/1/MERGA-2017-ChoyBH.pdf (accessed 20 May 2023).

Clapham, A. and Vickers, R. (2018) Neither a borrower nor a lender be: Exploring 'teaching for mastery' policy as borrowing, *Oxford Review of Education*, 44(6): 787–805. https://doi.org/10.1080/03054985.2018.1450745

Collopy, R. (2003) Curriculum materials as a professional development tool: How a mathematics textbook affected two teachers' learning, *The Elementary School Journal*, 103(3): 287–311. Available at: https://www.journals.uchicago.edu/doi/pdf/10.1086/499727 (accessed 6 September 2023).

Department for Education (2016) *South Asian method of teaching maths to be rolled out in schools*. Available at: https://www.gov.uk/government/news/south-asian-method-of-teaching-maths-to-be-rolled-out-in-schools (accessed 21 May 2023).

Diamond, J.B. (2007) Where the rubber meets the road: rethinking the connection between high-stakes testing policy and classroom instruction, *Sociology of Education*, 80(4): 285–313. https://doi.org/10.1177/003804070708000

Doig, B. and Groves, S. (2011) Japanese lesson study: teacher professional development through communities of inquiry, *Mathematics Teacher Education and Development*, 13(1): 77–93. Available at: https://files.eric.ed.gov/fulltext/EJ960950.pdf (accessed 18 May 2023).

Elliott, J.G. (2014) Lessons from abroad: Whatever happened to pedagogy?, *Comparative Education*, 50(1): 27–44. https://doi.org/10.1080/03050068.2013.871835

Fan, L., Zhu, Y. and Miao, Z. (2013) Textbook research in mathematics education: development status and directions, *ZDM Mathematics Education*, 45(5): 633–46. https://doi.org/10.1007/s11858-013-0539-x

Fan, L., Cheng, J., Xie, S. et al. (2021) Are textbooks facilitators or barriers for teachers' teaching and instructional change? An investigation of secondary mathematics teachers in Shanghai, China, *ZDM Mathematics Education*, 53(6): 1313–30. https://doi.org/10.1007/s11858-021-01306-6

Fujii T. (2018) Lesson Study and Teaching Mathematics Through Problem Solving: The Two Wheels of a Cart, in M. Quaresma, C. Winslów, S. Clivaz, J. da Ponte, A. Ní Shúilleabháin, and A. Takahashi (eds) *Mathematics lesson study around the world*. ICME-13 Monographs. Cham, Switzerland: Springer. https://doi.org/10.1007/978-3-319-75696-7_1

Groves, S. (2013) Implementing the Japanese problem-solving lesson structure, in V. Steinle, L. Ball, & C. Bardini (eds), *Mathematics education: yesterday, today and tomorrow: proceedings of the 36th annual conference of the Mathematics Education Research Group of Australasia* (pp. 712–714). Melbourne: MERGA. Available at: https://files.eric.ed.gov/fulltext/ED572909.pdf (accessed 18 May 2023).

Gu, F., Huang, R. and Gu, L. (2017) Theory and development of teaching through variation in mathematics in China, in R. Huang and Y. Li (eds) *Teaching and learning mathematics through variation*, pp. 13–41. Lieden, The Netherlands: Brill Sense.

Huang, R. and Bao, J. (2006) Towards a model for teacher professional development in China: Introducing Keli, *Journal of Mathematics Teacher Education*, 9(3): 279–98. https://doi.org/10.1007/s10857-006-9002-z

Huang, R. and Leung, F. (2017) Teaching geometrical concepts through variation: A case study of a Shanghai lesson, in R. Huang and Y. Li (eds) *Teaching and learning mathematics through variation*, pp. 151–168. Lieden, The Netherlands: Brill Sense.

Huang, R. and Li, Y. (2009) Pursuing excellence in mathematics classroom instruction through exemplary lesson development in China: a case study, *ZDM Mathematics Education*, 41: 297–309. https://doi.org/ 10.1007/s11858-008-0165-1

Huang, R., Su, H. and Xu, S. (2014) Developing teachers' and teaching researchers' professional competence in mathematics through Chinese Lesson Study, *ZDM Mathematics Education*, 46(2): 239–51. https://doi.org/10.1007/s11858-013-0557-8

Huang, X., Huang, R., Huang, Y. et al. (2019) Lesson study and its role in the implementation of curriculum reform in China, in R. Huang, A. Takahashi and J. da Ponte (eds) *Theory and Practice of Lesson Study in Mathematics*, pp. 229–52. Cham, Switzerland: Springer. https://doi.org/10.1007/978-3-030-04031-4_12

Isoda, M. (2007) *Japanese lesson study in mathematics: its impact, diversity and potential for educational improvement*. Singapore: World Scientific.

Ivars, P., Fernández, C., Llinares, S. and Choy, B.H. (2018) Enhancing noticing: Using a hypothetical learning trajectory to improve pre-service primary teachers' professional discourse, *Eurasia Journal of Mathematics, Science and Technology Education*, 14(11): 1–16. https://doi.org/10.29333/ejmste/93421

John, P. (1991) A qualitative study of British student teachers' lesson planning perspectives, *Journal of Education for Teaching*, 17(3): 301–20. https://doi.org/10.1080/0260747910170307

John, P. (2006) Lesson planning and the student teacher: re-thinking the dominant model, *Journal of Curriculum Studies*, 38(4): 483–98. https://doi.org/10.1080/00220270500363620

Krathwohl, D.R. (2002) A revision of Bloom's taxonomy: an overview, *Theory into practice*, 41(4): 212–18. https://doi.org/10.1207/s15430421tip4104_2

Kullberg, A., Runesson Kempe, U. and Marton, F. (2017) What is made possible to learn when using the variation theory of learning in teaching mathematics?, *ZDM Mathematics Education*, 49: 559–69. https://doi.org/10.1007/s11858-017-0858-4

Leong, Y.H., Ho, W.K. and Cheng, L.P. (2015) Concrete-Pictorial-Abstract: Surveying its origins and charting its future, *The Mathematics Educator*, 16(1): 1–18. Available at: https://repository.nie.edu.sg/bitstream/10497/18889/1/TME-16-1-1.pdf (accessed 6 September 2023).

Lewis, C. (2002) Does lesson study have a future in the United States?, *Nagoya journal of education and human development*, January 2002(1): 1–23. Available at: https://files.eric.ed.gov/fulltext/ED472163.pdf (accessed 6 September 2023).

Lewis, C. (2016) How does lesson study improve mathematics instruction?, *ZDM Mathematics Education*, 48: 571–80. https://doi.org/10.1007/s11858-016-0792-x

Loughran, J. (2019) Pedagogical reasoning: The foundation of the professional knowledge of teaching, *Teachers and Teaching*, 25(5): 523–35. https://doi.org/10.1080/13540602.2019.1633294

Lowrie, T. and Patahuddin, S.M. (2015) ELPSA as a Lesson Design Framework, *Indonesian Mathematical Society Journal on Mathematics Education*, 6(2): 1–15. Available at: https://files.eric.ed.gov/fulltext/EJ1079522.pdf (accessed 6 September 2023).

Lowrie, T., Logan, T. and Patahuddin, S.M. (2018) A learning design for developing mathematics understanding: The ELPSA framework, *Australian Mathematics Teacher*, 74(4): 26–31. Available at: https://files.eric.ed.gov/fulltext/EJ1231012.pdf (accessed 12 May 2023).

Mansfield, J. and Loughran, J. (2018) Pedagogical equilibrium as a lens for understanding teaching about teaching, *Studying Teacher Education*, 14(3): 246–57. https://doi.org/10.1080/17425964.2018.1541274

Marton, F. (2015) *Necessary conditions of learning*. New York: Routledge.

Mathematics Mastery (n.d.) *Mathematics Mastery Secondary*. Available at: https://www.arkcurriculumplus.org.uk/our-programmes/secondary/mathematics-mastery (accessed 12 May 2023).

Matić, L.J. (2019) The teacher as a lesson designer, *Center for Educational Policy Studies Journal*, 9(2): 139–60. https://doi.org/10.26529/cepsj.722

Mendes, F., Brocardo, J. and Oliveira, H. (2021) Building opportunities for learning multiplication, in M. Isoda and R. Olfos (eds) *Teaching multiplication with lesson study: Japanese and Ibero-American theories for international mathematics education*, pp. 241–64. London: Springer Nature. https://doi.org/10.1007/978-3-030-28561-6

Mouratidis, A., Michou, A., Demircioğlu, A.N. and Sayil, M. (2018) Different goals, different pathways to success: Performance-approach goals as direct and mastery-approach goals as indirect predictors of grades in mathematics, *Learning and Individual Differences*, 61: 127–35. https://doi.org/10.1016/j.lindif.2017.11.017

Naroth, C. and Luneta, K. (2015) Implementing the Singapore mathematics curriculum in South Africa: Experiences of foundation phase teachers, *African Journal of Research in Mathematics, Science and Technology Education*, 19(3): 267–77. Available at: https://hdl.handle.net/10520/EJC181277 (accessed 6 September 6 2023).

NCETM (2017) *Five bid ideas in teaching for mastery*. Available at: https://www.ncetm.org.uk/teaching-for-mastery/mastery-explained/five-big-ideas-in-teaching-for-mastery/ (accessed 2 June 2023).

OECD (2010) *PISA 2009 Results: What students know and can do – student performance in reading, mathematics and science (Volume I)*. Available at: https://www.ncetm.org.uk/teaching-for-mastery/mastery-explained/five-big-ideas-in-teaching-for-mastery/ (accessed 2 June 2023).

Pang, J. (2016) Improving mathematics instruction and supporting teacher learning in Korea through lesson study using five practices, *ZDM Mathematics Education*, 48: 471–83. https://doi.org/10.1007/s11858-016-0768-x

Peng, M., Bao., J. and Ki, W. (2017) Bianshi and the variation theory of learning, in R. Huang and Y. Li (eds) *Teaching and learning mathematics through variation*, pp. 43–67. Lieden, The Netherlands: Brill Sense.

Piaget, J. (1952) *The origins of intelligence*. New York: International University Press.

Piaget, J. (1977) *The development of thought: Equilibration of cognitive structures*. New York: Viking.

Piaget, J. (1985) *The equilibrium of cognitive structures*. Chicago: University of Chicago Press.

Salimi, M., Suhartono, S., Hidayah, R. and Fajari, L.E.W. (2020) Improving mathematics learning of geometry through the concrete-pictorial-abstract (CPA) approach: collaborative action research, *Journal of Physics: Conference Series*, 1663(1): 1–8.

Scheiner, T., Montes, M.A., Godino, J.D. et al. (2019) What makes mathematics teacher knowledge specialized? Offering alternative views, *International Journal of Science and Mathematics Education*, 17(1): 153–72. https://doi.org/10.1007/s10763-017-9859-6

Seleznyov, S. (2018) Lesson study: an exploration of its translation beyond Japan, *International Journal for Lesson and Learning Studies*, 7(3): 217–29. https://doi.org/10.1108/IJLLS-04-2018-0020

Shimizu, Y. (2008) Exploring Japanese teachers' conception of mathematics lesson structure: similarities and differences between pre-service and in-service teachers' lesson plans, *ZDM Mathematics Education*, 40(6): 941–50. https://doi.org/10.1007/s11858-008-0152-6

Shulman L.S. (1987) Knowledge and teaching: foundations of the new reform, *Harvard Educational Review*, 57(1): 1–22. https://doi.org/10.17763/haer.57.1.j463w79r56455411

Simpson, A. and Wang, Y. (2022) Making sense of 'mastery': understanding of a policy term among a sample of teachers in England, *International Journal of Science and Mathematics Education*, 21(2): 581–600. https://doi.org/10.1007/s10763-021-10178-x

Sun, X. (2011) "Variation problems" and their roles in the topic of fraction division in Chinese mathematics textbook examples, *Educational Studies in Mathematics*, 76: 65–85. https://doi.org/10.1007/s10649-010-9263-4

Tall, D. (2008) Using Japanese lesson study in teaching mathematics, *Scottish Mathematical Council Journal*, 38: 45–50. Available at: https://homepages.warwick.ac.uk/staff/David.Tall/pdfs/dot2008d-lesson-study.pdf (accessed 6 September 2023).

Tíjaro-Rojas, R., Arce-Trigatti, A., Cupp, J. et al. (2016) A systematic and integrative sequence approach (SISA) for mastery learning: anchoring Bloom's revised taxonomy to student learning, *Education for Chemical Engineers*, 17: 31–43. https://doi.org/10.1016/j.ece.2016.06.001

Torff, B. and Sternberg, R.J. (2001) Intuitive conceptions among learners and teachers, in B. Torff and R.J. Sternberg (eds) *Understanding and teaching the intuitive mind*, pp. 15–38. New York: Routledge.

Tummons, J. (2014) Professional standards in teacher education: tracing discourses of professionalism through the analysis of textbooks, *Research in Post-Compulsory Education*, 19(4): 417–32. https://doi.org/10.1080/13596748.2014.955634

Wang, Y. and Brown, C. (2021) Improving mathematical reasoning – the professional development challenge, *Professional Development Today*, 22(4). Available at: https://www.teachingtimes.com/improving-mathematical-reasoning-the-professional-development-challenge/ (accessed 6 September 2023).

Watson, A. and Ohtani, M. (eds). (2015) Task design in mathematics education: An ICMI study. Berlin: Springer.

Webb, M. E. (2002) Pedagogical reasoning: issues and solutions for the teaching and learning of ICT in secondary schools, *Education and Information Technologies*, 7(3): 237–55. https://doi.org/10.1023/A:1020811614282

Yuan, H. and Huang, X. (2019) China-England mathematics teacher exchange and its impact, *Frontiers of Education in China*, 14: 480–508. https://doi.org/10.1007/s11516-019-0023-7

Zhang, J., Wang. R. and Kimmins, D. (2017) Strategies for using variation tasks selected mathematics textbooks in China, in R. Huang and Y. Li (eds) *Teaching and learning mathematics through variation*, pp. 213–40. Lieden, The Netherlands: Brill Sense.

4 The Causal Connectivity Framework – Embedded reasoning in lesson planning

Key arguments

In this chapter, we introduce a pedagogical framework called the Causal Connectivity Framework (CCF). The CCF promotes integrating reasoning into teachers' choices of classroom activities. It is about generating learning experiences through series of connected activities that allow students to recognize their own learning process as they reflect on their learning experience. Lessons aim to build up students' narratives around mathematics knowledge. Once this process is in place, learning outcomes will follow automatically. The basis of lesson planning, therefore, becomes reasoning on how to unfold a certain topic in order to drive the development of students' understanding.

Key terms

- Coherence
  Coherence in lesson planning means the presence of a central theme that is developed through various activities. Activities are interconnected, leading to the emergence of a synthesis of mathematical knowledge. Coherence of this kind allows students to develop a 'new' comprehensive system for understanding mathematics.

- Causal connectivity
  One way to achieve coherence is through reasoning upon the connections between chosen activities. These connections should have causal flow, building up a logical discourse which is made explicit for students to reflect on.

The CCF for reasoning embedded in lesson planning was first outlined in the leading UK professional teaching journal *Mathematics Teaching* (Dawson and Wang 2019). This chapter expands on that work with a full theoretical discussion of the framework. The overarching purpose is to demonstrate a possible lesson structure, or a supportive teaching trajectory, and to establish a lesson routine that elaborates the perspective of promoting reasoning. The aim is to show how mathematical reasoning can be supported through sequencing carefully selected examples, while developing conceptual understanding. Then, the chapter zooms into reflections from the authors to highlight some of the advantages that the CCF may afford.

4.1 Coherence

The framework mainly sits within a step-based, structure-oriented lesson planning approach (see chapter 3). It emphasizes planning the coherence of the lesson(s), based on the principle that mathematics topics are interconnected in nature. The term 'coherence' was used by the TIMSS Video Study to gauge the quality of classroom teaching (Leung 2005). Coherence was one of four dimensions, with the other three being presentation, engagement and overall quality. In the 1999 report on this Video Study, 'coherence' referred to the 'interrelation of all mathematical components of the lesson', both implicit and explicit (Hiebert 2003, p. 196). Assessing coherence means judging the degree to which themes appear in a lesson or lessons, in ways ranging from the appearance of multiple unrelated themes (rated 1 out of 5) to the existence of a central theme that progresses saliently through the whole lesson (rated 5 out of 5). A striking characteristic of East Asian classrooms in this study is that they featured greater coherence; 90 per cent of Hong Kong lessons, for example, showed thematical coherence.

In addition, East Asian classrooms were more advanced in content, which means they were more proof-related and concerned to the greatest extent with the use of mathematical language. They were also more fully developed in their presentations, emphasizing sound mathematical reasoning (Leung 2005). In Japan, the lesson was viewed as a unit through which to form a coherent story, while US lessons consisted of smaller independent units (Clarke, Keitel and Shimizu 2006). When comparing the seated work time spent in lessons, students in Japan experienced more tasks concerned with inventing new solutions, playing up thinking or reasoning, than tasks aimed at practising routine procedures. The latter were the main focus of lessons in Germany and the United States (Shimizu 2009). From these findings, we see the importance of the lesson planning stage as the first step in building plausible coherence, especially as it relates to carefully selecting activities within a topic. These activities are not selected mainly with consideration for engagement or differentiation, but for how they become inextricably intertwined within the flow, connecting pieces of knowledge through reasoning.

The interconnected nature of mathematics topics, however, provides the foundation for this coherence. The topics of division, fraction and ratio are a good example. If we were to separate the symbol for division ÷ into $\frac{-}{-}$ and :, we would be left with a fitting expression of division's place between fraction and ratio. Then, the instrumental understanding of simplifying fractions and ratios becomes relational understanding. A connectedness between parts arises in students' web of knowledge. Hence, we see a remarkable synthesis of perspectives on division, highlighting the topic's nature as connected to other ones both procedurally and conceptually. These connections eventually lead us to reflect on our lesson planning and synthesize our approach to deepening understanding of mathematics topics in the curriculum.

The diagram in Figure 4.1 initially developed by Dawson, shows connections between knowledge about sequences as this knowledge develops from lower-secondary to higher-secondary levels. In this way, sequences are linked with the concepts of function, equations, shapes, etc. The birth of this diagram was not an easy one, partly because of how the secondary curriculum structures exposure to all concepts in a vertical way – dividing it between several subject contents. This is common practice for curricula around the world. The national curriculum includes six relevant topic areas: number; algebra; ratio; proportion and rates of change; geometry and measures; probability; and statistics (DfE 2021), while Singapore's curriculum has

Figure 4.1 The Causal Connectivity Framework

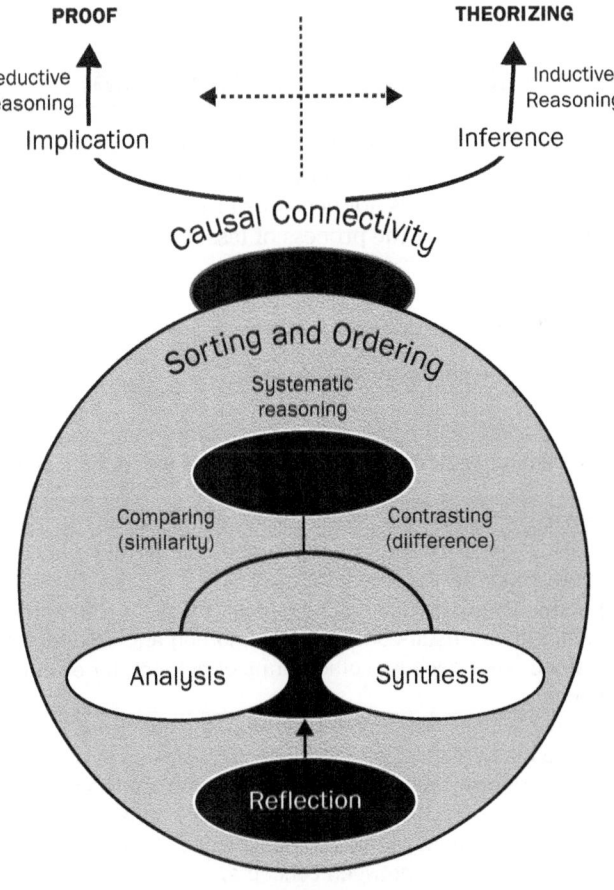

three main areas: number and algebra; geometry and measurement; statistics and probability (Ministry of Education (MoE) 2019). However, in practice, we have been seeing how schools in north-east England are revising their schemes of work so that topics are linked in ways that fit with different ability groups.

The aims of the national curriculum emphasize reasoning, conceptual understanding and fluency (DfE 2021), while Singapore's particularly values problem-solving (MoE 2019) and pursues this through a horizontal approach. The knitting together of the three aims of the national curriculum directly drives lesson planning, while Singapore's typical lesson structure builds from the aim of solving a problem. The purpose of the CCF, on the other hand, is to gear lesson planning towards mathematical reasoning. By using this framework, we promote mathematical activities of the kinds described by Schoenfeld (1994, p. 68): 'looking into perceived structure, seeing connections, capturing patterns symbolically, conjecturing and proving, and abstracting and generalizing'. The ultimate aim is to find balance between two interpretations: mathematics as cognitive process and mathematic as social processes (Wood, Cobb and Yackel 1995).

4.2 The Causal Connectivity Framework

The framework has five interconnected phases (see Figure 4.1): (1) relevance; (2) analysis and synthesis; (3) sorting and ordering; (4) causal connectivity; and (5) proof or theorizing. It is not necessary for a lesson to start from phase 1 and end at phase 5, either in the process of lesson planning or in the teaching trajectory itself.

Box 4.1 A short explanation of the CCF

- Relevance: displaying a balance between affective and cognitive aspects of maths learning, a starting point for building a mathematical narrative in a lesson.
- Analysis and synthesis: merging different approaches to the topic, a move from empirical generalization to structural generalization, acknowledging the ultimate importance of the latter.
- Sorting and ordering: articulating rules used for sorting and ordering, and the presentation of results based on this sorting and ordering; the ability to understand other rules and other forms of product, for example using a CPA approach.
- Causal connectivity: connectedness within mathematics itself, and with contexts outside mathematics, establishing interconnections with causal flow.
- Proof or theorizing – proof via implication or theorizing via inference: decontexutalizing in-context mathematical tasks, 'new' conceptual systems, a hierarchical relationship between pieces of knowledge, or a reconstruction of the relations between concepts.

4.2.1 Phase 1: Relevance

How would the teacher go about selecting the first activity of the class? In other words, what are the first five minutes of the class there to achieve? At this point in the lesson, more consideration can be placed on affective rather than cognitive aspects (McGrath 1992). In UK schools, this might mean attending to behaviour management, settling students down and doing the register for the purposes of safeguarding. Within problem-solving lessons in mind, on the other hand, things start with students working individually to solve a problem (Takahashi 2006). In fact, the relevance phase of the CCF recognizes the need to strike a balance between broadly fostering students' engagement in the mathematics learning ahead and more specifically motivating them and orienting them towards particular pre-planned learning objectives (such as through setting problem-solving tasks).

Achievement of this balance could take the form of engaging students in various applications of mathematics. The activity chosen or developed by teachers based on their understanding of the class should take students back to the heart of the discipline of mathematics, whether implicitly or explicitly. Their experience in this phase should imbue a sense that students have a relationship with the particular mathematics content as well as with the ethos of the classroom. This comes through dialogue, where discussion and the use of representations and mathematical language are encouraged from the very first moment.

This opening activity could be based on inquiry that raises a certain degree of perplexity and cognitive conflict, with the aim of motivating students to search for a solution. But such an activity also has to inspire in students an awareness of the limitations of their previous understanding in mathematics. It could also take a playful character, promoting social interaction, to encourage students to appreciate the beauty, utility and value of mathematics. But such an activity also has to lead to critical reflections on the process of abstracting mathematical structure from the specifics of the problem situation. For example, *odd-one-out* activities often highlight the key conceptual understanding that can arise when students are driven by their instinct to attempt reasoning associated with mathematics topics.

Thus, the activity chosen for the relevance phase enables students to become deeply and actively involved in the constructions of mathematics significance – the learning objectives – by creating their own mathematical narratives. These narratives prepare the ground for a structural transformation of students' web of mathematical knowledge and understanding. The next chapter will introduce a way of revealing the mathematical structure of probability through the drawing of a monster, while chapter 6 illustrates an approach to the concept of trigonometry linked to similar triangles, which itself links to the concept of scale factor, an example where conceptual understanding is expanded as students match sides to the angle.

4.2.2 Phase 2: Analysis and synthesis

During the analysis and synthesis phase, the activities chosen are the first steps in procedural variation, and a starting point for moving from empirical generalization to structural generalization. The analysis focuses on decontextualized aspects. The analysis may not be systematic at this stage, but rather this phase is concerned with conceptualizing the construction of mathematics understanding. Here, different types of presentation, such as verbal, pictorial, literary, gestural or methodological methods, imply to students the rules for analysis, and they begin to introduce informal and unconscious problem-solving strategies. Through this phase of analysis, students organize or synthesize information, and it is important for teachers to allow student understanding or misunderstanding to take different routes as students first begin to elaborate their own solutions. This means teachers need to employ pedagogical reasoning during the planning stage, as they anticipate how their students will interact with the activities. The present phase lays the foundation for teachers to mediate their students' critical analysis of the problem at hand, and to reconstruct students' conceptual image of this piece of mathematical knowledge.

Thus, through reflection, students come to realize that empirical generalization is not satisfactory anymore; the alternative, moving to structural generalization, must now be developed. For hard-to-learn topics, such as solving quadratic equations, trigonometry and rationalizing denominators (at key stage 4), the analysis and synthesis phase involves modelling. The critical-reflective dimension of mathematical modelling is a process of elaboration, critical analysis and validation of a model, and all of this allows classroom dialogue to proceed in a rational way (Orey and Rosa 2017). The result is that reasoning starts to appear in the mathematics classroom, as related to the relevant topic. Sometimes it can be difficult to show the process of elaboration; one example is with the relationship between gradient and steepness in a linear graph. Certain pieces of mathematics software can be useful here; in the digital age, one benefit of using dynamic mathematics software as learning media is the ability to 'show the process of mathematical object generation' (Wijaya, Ying and Purnama 2020, p. 216).

This phase also brings collective mathematical sense-making into the classroom, as students begin to compare and justify differences and similarities in findings, harnessing awareness of the approaches of others to synthesize different exploratory paths and insights. Thus, generalization by exploration implicitly becomes an integral part of lesson planning. At the same time, some degree of standardization, an orientation of students towards the scientific approaches of mathematics, begins to emerge. This forms the basis for practice, ahead of further procedural variation to come. Different directions from which to approach new concepts are first reached by attempting to isolate underlying invariant structure or structural elements, and this is supported by reflection.

4.2.3 Phase 3: Sorting and ordering

Sorting and ordering are competencies in logic, training in which starts from pre-school times with participation in number skill activities such as counting (Clements 1984). Activities in the sorting and ordering phase of the CCF offer opportunity for students to develop their own ways of dealing with a problem. There are three aspects: the presentation of results as the product of sorting and ordering; the ability to articulate rules used for sorting and ordering; and the ability to understand other rules and other forms of product. During this process, activities should be chosen to help students in identifying and critiquing the relevant mathematical structure. The activities should also encourage students to consider the relative value of representations that may prioritize one aspect over another.

There is another crucially important issue to explore here: mathematical representations. We take the definition of mathematical representations from Goldin (2020), viewing them as products, either visible or tangible, and the processes of producing representations. If viewed as products, representations are tools to present ideas mathematically, for example in graphical, tabular, verbal or symbolic ways. However, using different representations to explain the same concept does not straightforwardly lead to deeper understanding, at least not for sixth-grade Cypriot students (Gagatsis, Elia and Kyriakides 2003). Therefore, activities chosen for this phase should emphasize the processes of producing or constructing representations, and they should encourage evaluation of the products, rather than merely using representations for the sake of it.

Use of certain patterns, such as a concrete-pictorial-abstract (CPA) strategy, is encouraged here. Activities in the sorting and ordering phase invite exploration through different representations. The next chapter will give an example from this phase of the benefits of moving from concrete to pictorial representation, and from pictorial to abstract representation in solving problems. Vergnaud (1998, p. 174) pointed out that representation 'enables us to anticipate future events'. His notion of representation depends on the role of language and symbols. Therefore, building activities here means going beyond a focus on depth of understanding, and it instead requires emphasis on directions for exploring the topic – all leading towards the causal connectivity phase.

Structural generalization comes in during sorting and ordering. Activities here might not touch on all possible types of representation to build up the meaning of the new concept in its totality. But structurally, there has to be enough to distinguish this topic from others and to extract the invariables. Then, reflection on the series of activities undertaken so far makes students aware of how they learn and of how mathematics knowledge unfolds through these activities. The aim is to lay the foundations for students to cultivate a habit of systematic and deliberate reflection.

4.2.4 Phase 4: Causal connectivity

The causal connectivity phase has two aspects: connectedness, and causal flow from activities in the previous phases. Hiebert and Carpenter (1992) outlined the idea that making connections in mathematics is essential for understanding development. In other words, how strongly and richly students make connections is related to how deep their understanding is. From the perspective of understanding, there are two aspects of connectedness or making connections that are generally encouraged:

1   Connections among mathematical ideas and domains, such as a set of related topics or multiple mathematical representations.
2   Connections with contexts outside mathematics, such as real-life situations or issues in other subject areas. In some schools, cross-curriculum links have become an explicit part of the lesson planning stage. The website of Sedgefield Community College (n.d.), for example, lays out a selection of cross-curricular themes, including scientific discovery and additional skills such as teamwork.

Activities chosen for this phase are geared towards deepening understanding of the relevant topic in the learner's web of mathematics knowledge. But causal connectivity means more than that; the framework is based on developing reasoning through understanding.

From a pedagogical perspective, making connections appears as one of the big ideas of teaching for mastery (NCETM 2017). However, there is no universally agreed way for how students are supposed to make connections (Mills et al. 2009). If in lesson planning the connection is implicit, then in delivery it should become explicit. Teachers in Shanghai build the connections between tasks, moving from implicit to explicit levels (Mok and Kaur 2006), through the variation/*bianshi* 变式 approach (see chapter 3, section 3.2.2). Through variation, students learn to classify what is invariable so that the key points of the mathematics concept are understood. Through the connectedness of these activities, the critical points occur with procedural variation. Based on the understanding of 'mathematics as a network of layers of interconnected nodes' (Pepin 2021, p. 1223), the focus of this phase is to make these connections explicit in planning so that they could be highlighted in the delivery stage to show the nodes of interconnection.

Even more importantly, this connectedness also has causal flow, a direction. This is what encourages students to wonder and anticipate what comes next. Causal flow appears in the form of either generalization, justification or proof (the three types of reasoning in a mathematical way: see chapter 1, section 1.2.1), each of which follows when opportunities are created for students to reflect. Kaur (2012, p. 75) proposed a 'what...' strategy with, for example, 'what if' questions becoming a basic building block of logical discourse. It is especially important for instructions and guidance related to these activities to be considered fully at the lesson planning stage. This phase is the point where there is a shift from reasoning within activities towards modelling the formal mathematical activity.

4.2.5 Phase 5: Proof or theorizing

After the causal connectivity phase, two types of activities form the last phase of deepening students' understanding: proof via implication and theorizing via inference. Proof here refers to the application of concepts, including in real-life scenarios or linking with other subject areas, so that students can test out the mathematical ideas in context. Jablonka and Bergsten (2010) advocated the idea that authentic tasks should have two elements: comprehensiveness and fidelity. Comprehensiveness means taking problems that fit with both in-school and out-of-school situations, while the fidelity of a task concerns the use of appropriate representations for possible solution strategies. In another words, this is a stage of decontextualization of previously in-context mathematical tasks. It allows students to develop their analytical skills as based on the 'ecology' of mathematical knowledge derived from the causal connectivity phase. Furthermore, Stylianides and Stylianides (2008) advised that higher fidelity tasks with high cognitive demands should be built into a lesson by gradually increasing the level of challenge from previous tasks.

On the other hand, theorizing aims to test out the 'new' conceptual system established in the previous stage. This conceptual system might take the form of a hierarchical relationship between pieces of knowledge (in a vertical arrangement), or of a reconstruction of the relations between concepts (in a horizontal way). Tasks chosen tasks here direct subsequent action and have more specific purposes, allowing students to demonstrate their ability to make the concepts or mathematical ideas more visible, especially for those that students could not grasp without the mediation of causal connectivity. Therefore, tasks offer students chance to demonstrate the internal organizational systems of relevant mathematical ideas.

This phase also exemplifies the final level of mathematical understanding highlighted in various theories of mathematical understanding, such as APOS theory and Pirie and Kieren's model (Pirie and Kieren 1994). APOS – action, process, object, and schema (Dubinsky and McDonald 2002) – describes processes of understanding for abstract concepts. Its last stage, schema, is related to the 'concept image' that students have developed, something which occurs after the causal connectivity phase in our framework.

4.3 Reflection on the Causal Connectivity Framework

The phases in the CCF are visited with students on the way to developing an understanding of the particular mathematics concept being studied. In this section, we will first re-examine the assumptions enthroned in this framework: a map of observable processes in teachers' own narrative of logic and reasoning. Once this foundation is established, reflection on the reasoning process becomes a necessary response. We will look next at reflection, and then finally consider how the role of the teacher might be rethought in the context of reasoning-based lesson planning.

4.3.1 Reflection 1: An interpretation of learning experience

The CCF stems from a belief that learning experience counts, and it is an application of social phenomenology: not only the learning experience, but more importantly the interpretation of this learning experience constructs the reality (Fermín-Gonzáles and Domínguez-Garrido 2021). To construct this reality, in terms of the understanding of mathematical topics, lesson planning should not focus primarily on a set of check-list items and the ultimate aim of students being able answer a certain question by the end of the lesson. Rather, it is about a recognition of the learning process, one in which all parts are interdependent and combine to form a narrative for the lesson.

This narrative has phases, while the framework itself is a way of describing processes observable to anyone inclined to account for, and reflect on, what they do mathematically. The CCF provides an opportunity to explain the reasoning behind lesson planning. The lesson planning stage needs to identify the basic concepts to construct the new topic (relevance), analysing the understanding required and subsequently synthesizing it via a learning sequence (analysis and synthesis), sorting and ordering the concepts in a manner which presents a clear systematic developmental pathway leading to new understanding (causal connectivity). This constitutes the causal connectivity of the framework's name.

Bringing the CCF into the planning and structuring of classroom activities can facilitate the natural development of reasoning for students. It offers a means for devising learning sequences – sequences of connected activities – as the basis for nurturing and facilitating the development of these skills. The framework harnesses different types of reasoning, so structuring lessons with these processes in mind cements and develops these reasoning types with mathematical skills, while promoting and strengthening the value attached to them in the minds of participants. As such, it approaches reasoning not as something to be taught, but rather as the focus of a carefully designed learning environment – the sum of the activities with key questions prepared at the lesson planning stage. Students' hypothetical learning trajectory is structured in a coherent way.

4.3.2 Reflection 2: About reflection

Reflection, as one of several metacognitive activities, is a measure to monitor one's own work, either in assessing the procedures and methods used to solve a problem, or in evaluating the relevance of a concept in the broader mathematics system (Kaune 2006). In other words, reflection is this process of mind looking at mind, presenting itself to itself as if standing in front of a mirror. Through the CCF, attention is trained towards the cognitive process via reflection.

Reflection is slightly different from teaching metacognitive strategies. Metacognition is about justifying the evaluation of one's own knowledge (Son, Furlonge and Agarwal 2020). It paves the way for the emergence of a feedback mechanism, being either conscious or automatic, and either general or domain-specific (Veenman et al. 2006). These are associated with how students plan, monitor and evaluate their learning. Suggestions include explicitly teaching the strategies to students through seven steps: 'activating prior knowledge, explicit strategy instruction, modelling of learned strategy, memorization of strategy, guided practice, independent practice, and structured reflection' (Education Endowment Foundation (EEF) 2021, p. 14). Reflection in the CCF, however, occurs at each phase, so that there is an intelligible sense of how activities cumulatively fit together, and so students fully apprehend the coherence of the lesson.

4.3.3 Reflection 3: Rethinking the role of the teacher

The nature of activities selected for a lesson has implications for classroom instruction as well as for the role of the teacher. The role of the teacher becomes no longer to explain concepts and procedures in increasing detail, modelling step-by-step. Instead, it is the task of the student to come up with explanations based on previous activities, to attribute meaning in the student's own schema, and to develop their web of knowledge in mathematics as well as their understanding of links to other concepts. This is the fundamental pedagogical principle and intention behind implementation of the CCF. The teacher's role within the active classroom is to promote reflection and probe understanding, guiding students through the process using structured plenaries and careful questioning.

This reflects Ernest's (1989) conclusions about the teacher's role in facilitating intended outcomes; the teacher at once acts as an instructor to emphasize skill mastery, an explainer to support conceptual understanding and a facilitator to develop problem-solving. At the classroom level, the facilitator role is predominant, as the activities themselves feed the learning process, nourishing students' understanding or alleviating their mathematics anxiety. The teacher carefully plans learning sequences to implement within the active classroom so that, upon reflection, students have a sufficient evidence base (experiential or otherwise) from which to ascertain implications, draw inferences or make conclusions. This naturally promotes and develops the skills and processes of reasoning. Once students have had sufficiently prolonged exposure to this pedagogical method, a metacognitive level of understanding can be generated. Simply by directing them to reflect upon how their lessons have taken shape, this draws out the structure and principles at play and constitutes reflection on their own processes of reasoning.

4.4 What comes next?

In this chapter, we have discussed the importance of coherence in creating quality mathematics lessons at the observable level. Coherence is critically dependent on how well themes in the lesson are interconnected. In turn, reasoning affords this interconnection. We proposed the CCF, with its five phases, to promote this coherence in the context of observed reasoning processes and reflections. In turn, the role of the teacher is adjusted towards facilitation.

The next two chapters offer two examples of how this all can play out in practice. One example, from the transition stage, concerns probability. The second, at later secondary stage, is on the topic of trigonometry. Both examples are taken from two projects we have carried out in north-east England's schools over the past several years. Both lesson plans continue to be rolled out on a local scale.

Practice activities

Think of two topics: probability at lower-secondary level and basic trigonometry. What other topics are they fundamentally linked to? Here is a concept diagram for the topic of sequences (see Figure 4.2); think of how to create your own narratives about probability (see Figure 4.3) and basic trigonometry (see Figure 4.4).

Figure 4.2 Narrative around the topic of sequences

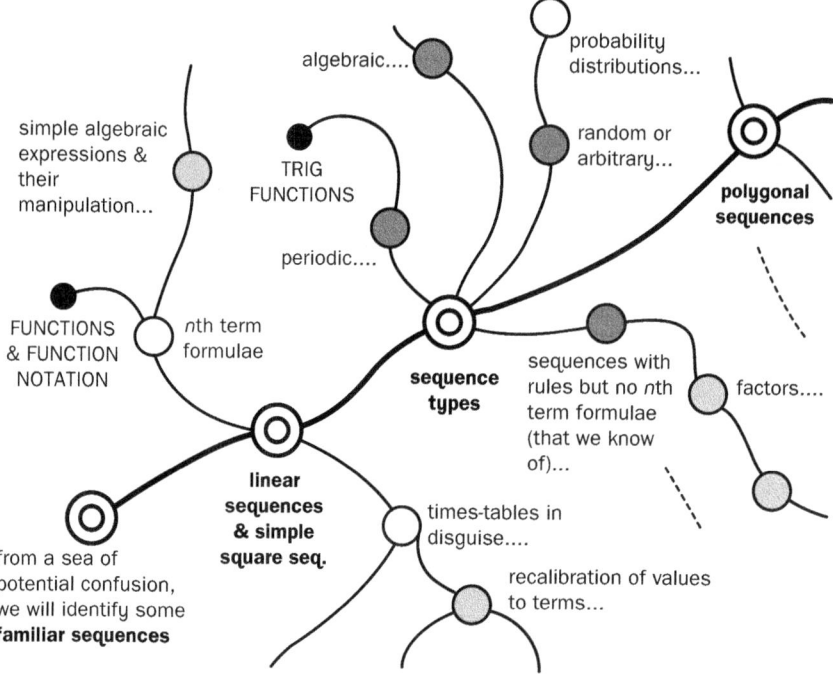

Figure 4.3 Your narrative about the topic of probability

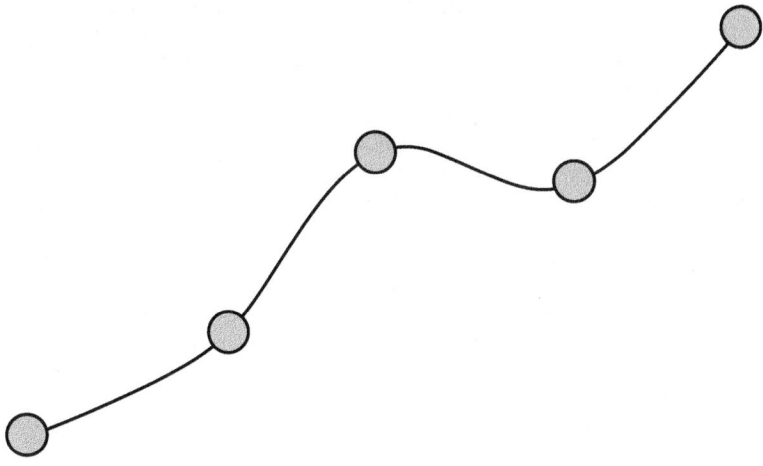

Figure 4.4 Your narrative about the topic of trigonometry

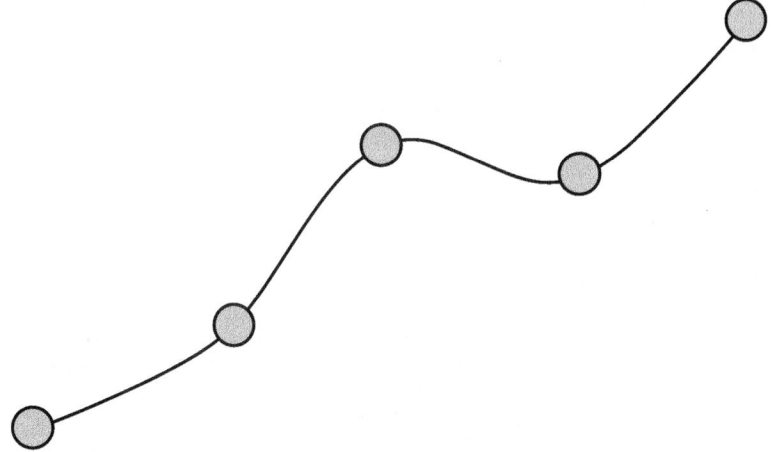

References

Clarke, D., Keitel, C. and Shimizu, Y. (2006) *Mathematics classrooms in twelve countries: The insider's perspective.* Lieden, The Netherlands: Brill.

Clements, D.H. (1984) Training effects on the development and generalization of Piagetian logical operations and knowledge of number, *Journal of Educational Psychology,* 76(5): 766–76. https://doi.org/10.1037/0022-0663.76.5.766

Dawson, J. and Wang, Y. (2019) How does a mastery lesson unfold? *Mathematics Teaching,* MT267: 11–14. Available at: https://durham-repository.worktribe.com/output/1290845 (accessed 6 September 2023).

Department for Education (2021) *National curriculum in England: mathematics programmes of study*. Available at: https://www.gov.uk/government/publications/national-curriculum-in-england-mathematics-programmes-of-study (accessed 12 July 2023).

Dubinsky, E. and McDonald, M. (2002) APOS: A constructivist theory of learning in undergraduate mathematics education research, in D. Holton (ed.) *The teaching and learning of mathematics at university level*, pp. 275–82. New York, NY: Kluwer.

Education Endowment Foundation (2021) Metacognition and self-regulated learning – guidance report. Available at: https://d2tic4wvo1iusb.cloudfront.net/eef-guidance-reports/metacognition/EEF_Metacognition_and_self-regulated_learning.pdf?v=1668185151 (accessed 20 June 2023).

Ernest, P. (1989) The impact of beliefs on the teaching of mathematics, *Mathematics Teaching: The State of the Art*, 249: p. 254. Available at: https://education.exeter.ac.uk/research/centres/stem/publications/pmej/impact.htm (accessed 5 June 2023).

Fermín-González, M. and Domínguez-Garrido, M.C. (2021) A phenomenological approach to Intercultural Initial Education, *Intercultural Education*, 32(6): 624–48. https://doi.org/10.1080/14675986.2021.1962251

Gagatsis, A., Elia, E. and Kyriakides, L. (2003) The nature of multiple representations in developing mathematical relationships, *PME CONFERENCE*, 1(Conf 27), p. 287–undefined.

Goldin, G.A. (2020) Mathematical representations, in S. Lerman (ed.) *Encyclopedia of mathematics education*. Cham, Switzerland: Springer.

Hiebert, J. (2003) Teaching mathematics in seven countries: results from the TIMSS 1999 Video Study. Washington, DC: DIANE Publishing.

Hiebert, J. and Carpenter, T.P. (1992) Learning and teaching with understanding, in D. A. Grouws (ed.) *Handbook of research on mathematics teaching and learning*, pp. 65–97. New York: Macmillan.

Jablonka, E. and Bergsten, C. (2010) Theorizing in mathematics education research: Differences in modes and quality, *Nordisk matematikkdidaktikk*, 15(1): 25–52. Available at: https://www.diva-portal.org/smash/get/diva2:983080/FULLTEXT01.pdf (accessed 6 September 2023).

Kaune, C. (2006) Reflection and metacognition in mathematics education – tools for the improvement of teaching quality, *Zentralblatt für Didaktik der Mathematik*, 38: 350–60. https://doi.org/10.1007/BF02652795

Kaur, B. (2012) Some "What" strategies that advance reasoning and communication in primary mathematics classrooms, in B. Kaur and T. Toh (eds.) *Reasoning, Communication And Connections In Mathematics: Yearbook 2012, Association of Mathematics Educators*, pp. 75–88. Singapore: World Scientific.

Leung, F.K.S. (2005) Some characteristics of East Asian mathematics classrooms based on data from the TIMSS 1999 video study. *Educational Studies in Mathematics*, 60(2): 199–215. https://doi.org/10.1007/s10649-005-3835-8

McGrath, I. (1992) Lesson Beginnings, *Edinburgh Working Papers in Linguistics*, 3: 90–108. Available at: https://files.eric.ed.gov/fulltext/ED353842.pdf (accessed 6 September 2023).

Mills, M., Goos, M., Keddie, A. et al. (2009) Productive pedagogies: A redefined methodology for analysing quality teacher practice, *The Australian Educational Researcher*, 36: 67–87. https://doi.org/10.1007/BF03216906

Ministry of Education, Singapore (2019) Mathematics syllabuses. Available at: https://www.moe.gov.sg/-/media/files/secondary/syllabuses/maths/2020-express_na-maths_syllabuses.?la=en&hash=E79043503E0EE64FA579D7514760663151459ED9 (accessed 1 June 2023).

Mok, I.A.C. and Kaur, B. (2006) 'Learning task' lesson events, in D. Clarke., J. Emanuelsson., E. Jablonka and I. Mok *Making connections: Comparing mathematics classrooms around the world*, pp. 147–63. Lieden, The Netherlands: Brill.

NCETM (2017) *Five bid ideas in teaching for mastery*. Available at: https://www.ncetm. org.uk/teaching-for-mastery/mastery-explained/five-big-ideas-in-teaching-for-mastery/ (accessed 5 June 2023).

Orey, D.C. and Rosa, M. (2017) Developing critical and reflective dimensions of mathematical modelling, in A. Chronaki (ed.) *Mathematics Education and Life at Times of Crisis, Proceedings of the 9th International Mathematics Education and Society Conference*, pp. 771–82. Volos, Greece: MES9.

Pepin, B. (2021) Connectivity in support of student co-design of innovative mathematics curriculum trajectories, *ZDM Mathematics Education*, 53: 1221–32. https://doi. org/10.1007/s11858-021-01297-4

Pirie, S.E.B. and Kieren, T. (1994) Growth in mathematical understanding: How can we characterize it and how can we represent it? *Educational Studies in Mathematics*, 26(2–3): 165–90. https://doi.org/10.1007/BF01273662

Schoenfeld, A.H. (1994) Reflections on doing and teaching mathematics, in A.H. Schoenfeld (ed.) *Mathematical thinking and problem solving*, pp. 53–70. New York and London: Routledge.

Sedgefield Community College (n.d.) *Inevitable progress*. Available at: https://sedgefield.laidlawschooltrust.co.uk/144/inevitable-progress (accessed 25 May 2023).

Shimizu, Y. (2009) Capturing the structure of Japanese mathematics lessons: Some findings of the international comparative studies, *Colección Digital Eudoxus*, 18. Available at: https://www.researchgate.net/profile/Yoshinori-Shimizu/publication/237204382_CAP-TURING_THE_STRUCTURE_OF_JAPANESE_MATHEMATICS_LESSONS_SOME_FINDINGS_OF_THE_INTERNATIONAL_COMPARATIVE_STUDIES/links/56f1f79f08aee9c94cfd7859/CAPTURING-THE-STRUCTURE-OF-JAPANESE-MATHEMATICS-LESSONS-SOME-FINDINGS-OF-THE-INTERNATIONAL-COMPARA-TIVE-STUDIES.pdf (accessed 6 September 2023).

Son, L., Furlonge, N. and Agarwal, P. (2020) Metacognition: how to improve students' reflections on learning. Available at: http://pdf.retrievalpractice.org/MetacognitionGuide.pdf (accessed 2 June 2023).

Stylianides, A.J. and Stylianides, G.J. (2008) Studying the classroom implementation of tasks: High-level mathematical tasks embedded in 'real-life' contexts, *Teaching and Teacher Education*, 24(4): 859–75. https://doi.org/10.1016/j.tate.2007.11.015

Takahashi, A. (2006) Characteristics of Japanese mathematics lessons, *Tsukuba journal of educational study in mathematics*, 25(1): 37–44. Available at: https://www.collectedny. org/wp-content/uploads/2017/10/Takahashi-Characteristics-of-Japanese-Mathematics-Lessons-2006-1.pdf (accessed 6 September 2023).

Veenman, M.V., Hout-Wolters, V., Bernadette, H.A.M. and Afflerbach, P. (2006) Metacognition and learning: Conceptual and methodological considerations, *Metacognition and learning*, 1(1): 3–14. https://doi.org/10.1007/s11409-006-6893-0

Vergnaud, G. (1998) A comprehensive theory of representation for mathematics education, *The Journal of Mathematical Behavior*, 17(2): 167–181. https://doi.org/10.1016/S0364-0213(99)80057-3

Wijaya, T.T., Ying, Z. and Purnama, A. (2020) Using hawgent dynamic mathematic software in teaching trigonometry, *International Journal of Emerging Technologies in Learning (IJET)*, 15(10): 215–22. Available at: https://www.learntechlib.org/p/217054/ (accessed 6 September 2023).

Wood, T., Cobb, P. and Yackel, E. (1995) Reflections on learning and teaching mathematics in elementary school, in L.P. Steffe. and J. Gale (eds) *Constructivism in education*, pp. 419–40. New York: Routledge. https://doi.org/10.4324/9780203052600

5 Towards probability

Key arguments

In this chapter, we offer an example of how the CCF can be used in supporting students' transition from primary to secondary levels. This example reveals a bottom-up approach to exploring mathematics. We argue that the transition period must do more to ensure curricular and pedagogical continuity across levels. We propose a visual conceptual structure as a connecting principle for mathematical knowledge.

Key terms

In the chapter to come, we introduce a series of hands-on classroom activities in which students are asked to draw the heads, bodies, and legs of monsters, and then to create new monsters by exchanging these three parts. The possible outcomes of monsters created by combining constituent parts are referred to as a 'sample space' in mathematics. And systematically working out the possible outcomes depends on what is called a 'visual conceptual structure' in the mathematics education field. In this case, we particularly emphasize the analysis and synthesis phase in the CCF, the step in which students are encouraged to reflect on their methods for analysing and synthesizing ideas.

- Within mathematics knowledge: sample space

  A sample space at lower-secondary level means a set of all the possible outcomes or results of an experiment. When tossing a coin, there are two possible outcomes: heads and tails. So the sample space can be presented as (H, T). This is the foundation of probability theory in mathematics.

- Within theories: visual conceptual structure

  Bar modelling, tree diagrams, prime factor trees, and function graphs all show conceptual structures visually. These visual conceptual structures become sources to establish mental structures about how the topics or concepts under consideration work (we also call this sort of mental structure a 'schema').

- Within the CCF: reflection on analysis and synthesis

  When exploring the sample space in monster activities (in this chapter), in the analysis and synthesis phase, students reflect on their methods of analysis and synthesis. Their current methods, manipulating the parts or guessing, may not work. Through this reflection, the need for a systematic method emerges as an essential step. This illustrates the process of reflection for generalizing.

The philosophy of mathematics has two basic approaches: bottom-up and top-down (Cellucci 2013; Ernest 2018). A bottom-up approach seeks to explain mathematics through reference to students' experiences and interactions, for example in the case of the play strategy advocated by the Scottish Government (2015). A top-down approach, on the other hand, explains mathematics through mathematics itself, which means relying on some unproven assumptions. This chapter and the next one present more details of these two philosophical approaches. Table 5.1 summarizes the different perspectives in connection with reasoning (see chapter 1, section 1.2).

Table 5.1 Summary of the present two chapters

	Chapter 5 Towards probability	Chapter 6 Basic trigonometry
Approaches to mathematics	Bottom-up, starting from specific examples, using play to generalize the rules	Top-down, introducing basic trigonometry through the concept of similar triangles
Reasoning in the way of doing mathematics	Focusing on generalization by explanation	Focusing on justification by mathematizing
Reasoning in the mathematics classroom	Laying the foundation for the mathematics structures of statistics – probability, exponential expressions and binomial coefficient formula	Associated with trigonometry

With this overview in mind, this chapter starts by looking at the transition from primary to secondary level in a general way, and in terms of the maths knowledge that is recommended at this stage. It focuses on a particular approach to mathematics knowledge – the use of visual conceptual structure. Following on from a discussion of tree diagrams, we present a narrative approach to probability, drawing on a part of the project carried out with a team from the Durham Local Authority. At that time, the Play Project was a series to inspire a *play* approach towards learning at higher primary level. We present two cycles from the Play Project aiming to support students through the transition stage by employing a detailed sequence of tasks undergirded by the CCF. This sequence of tasks transformed a game (a play approach) into something different: a world of mathematics, an exploration of mathematical structure and a space where the transformation from real life to the mathematical world, and from conceptual intervention to abstract mathematical structure, were increasingly interweaved. Behind this lesson planning is the intention to give the classroom environment a particular character, one inspiring the ambition needed for students to try, but one promoting enough humility for them to accept that there is a certain mathematical understand hidden beyond their grasp – initially at least.

5.1 The mathematics transition from primary to secondary level

The transition from primary to secondary level has become systematized, with many schools offering a programme of events and visits for students and their parents. Regarding mathematics in particular, there are three interlinked factors identified as the main impacts on students during this phase: student self-regulation, school and academic aspects and social factors (Kaur, McLoughlin and Grimes 2022). This conclusion was based on a systematic review of literature from 1990 to 2020 in the area of mathematics transition. 'Self-regulation' points to students' attitudes and the development of their identities, and within this area, mathematics anxiety has been particularly highlighted, especially for girls. Girls tend to have higher anxiety and lower self-efficacy than boys. Compared with Year 7 students, Year 8 students of both genders reported greater maths anxiety (Deieso and Fraser 2019). 'School and academic aspects' include teacher knowledge gaps and curriculum continuity, both of which become central to mathematics engagement. 'Social factors' are the importance of parental engagement and the home environment. These findings are useful in directing our attention towards the particular difficulties of this transition, and the question now becomes one of *how* – how to address them in secondary schools.

Approaching the higher-primary and lower-secondary levels as a comprehensive whole, the EEF published a guidance report on improving mathematics at key stages 2 and 3, recommending that a joined-up understanding of curriculum, teaching and learning should prevail across the two (Higgins et al. 2022). In some secondary schools in north-east England, there have been appointments of an experienced primary school teacher to help design the curriculum of Year 7 maths (the first year of secondary school) and to support, in particular, under-achieving students. This sheds light on several key areas of potential development. It exemplifies efforts to put in place a connecting principle, concerning pedagogy and curriculum. But it is equally important to build students' capacities to pursue mathematical exploration, those which are both 'intellectually challenging and resilience building' (Tytler et al. 2008). Now, we turn to the question of *what* – what does this sort of mathematical exploration look like?

The transition stage is also seen as an opportunity for students to gain new understanding of mathematics, thus growing their confidence (Darragh 2013). Mathematics knowledge has typically been divided into conceptual knowledge and procedural knowledge (Hiebert and Lefevre 1986). For example, knowing that a negative number multiplied by a negative number makes a positive number – this is procedural knowledge. Debate surrounding these two types of knowledge has focused on the instructional order, asking, for instance, whether moving from conceptual to procedural is better. The best conclusion is that it is bidirectional: the two support each other (Rittle-Johnson, Schneider and Star 2015).

School mathematical knowledge is further divided into three types by Ofsted (2021): declarative knowledge, including maths facts, vocabulary and symbols; procedural knowledge, such as methods for solving simultaneous equations;

and conditional knowledge, which is a combination of declarative and procedural knowledge transformed into strategies for problem-solving. Our question now becomes a *which* question – about which order these three kinds of knowledge should be organized into. Let's take another perspective, the use of images to shift approaches to knowledge. Here, we focus on images, such as tree diagrams and Venn diagrams, which contain visual conceptual structure that weaves together these different types of mathematical knowledge.

5.2 Visual conceptual structure

The mastery approach advocates CPA representations. Pictorial representation does not necessary constitute conceptual entities as well as operations, but Venn diagrams, tree diagrams and graphs representing functions do (Fischbein 1977). This means that they are more than pictorial representations. They show the process as well as the solution. So, these diagrams are considered examples of a heuristic model, which means that they are 'internally consistent and generative' (Fischbein 1977, p. 158).

Regarding process, visual conceptual structure directs students towards meaningful reasoning when solving non-routine problems. McGalliard and Wilson (2016) find that one possible cause for students reasoning only superficially is that students tend to focus on recalling prior knowledge and their learning experiences instead of reasoning about the current context. In the case of fractions, explanation of the part-whole relationship construct could be supported by using a tree diagram – a structure system (Giacomone, Beltran-Pellicer and Godino 2019) – to organize mathematical objects into a hierarchical sequence. The exercise of being systematic in use of tree diagrams becomes a source to establish supporting classification schema and combinatorial schema (Nunes et al. 2014). When it comes to solutions, all the different possible outcomes in a probability problem, called the sample space, is the basis for computing the probabilities for an event.

Tree diagrams are frequently used in aspects of probability including combinatorics and statistics. One typical error in combinatorics is a faulty interpretation of the tree diagram, and students also seem to have difficulties identifying common mathematical structures in different probabilistic problems (Batanero and Sanchez 2013). Therefore, the probability cycle in the Play Project drew upon use of tree diagrams and culminated in the development of systematic and logical discourse to form the very foundation of relevant concepts.

5.3 Narrative on probability

Table 5.2 outlines two consecutive series of lesson sequences (the monster series and the dice series). They feature a similar patten, one which involves hands-on tasks at the beginning, moving on to linking a variety of mathematical concepts via visual conceptual structures, and ultimately developing the ground for probability. The monster series was initially published as a brief example of using the CCF in mathematics teaching (Dawson and Wang 2019), and we extend it here.

Table 5.2 Narrative on probability

Monster Series

Activity	Causal Connectivity Framework	Mathematical concepts
Activity 1: Drawing monsters	Relevance Relevance to daily life, with no mathematics involved at this stage	N/a
Activity 2: Creating new monsters	Analysis and synthesis	Sample spaces
Activity 3: Explaining answers Modelling mathematically – a systematic method	Sorting and ordering	Combination, possibility, tree diagrams, power
Activity 4: Explaining the tree diagram	Causal connectivity with reflection	
Activity 5: Generalizing the findings	Theorizing	Algebraic expression of the power

Dice Series

Activity	Causal Connectivity Framework	Mathematical concepts
Activity 1: Throwing a pair of dice and adding the two numbers together – how many outcomes are possible?	Analysis and synthesis	Sample spaces, tree diagrams
Activity 2: Explaining answers in a systematic way	Sorting and ordering	The Cartesian coordinate system
Activity 3: Finding the most likely outcome and the least likely outcome	Causal connectivity	Probability, fraction, percentage

5.3.1 Monster series

Activity 1: Drawing monsters (with a group of three students) Imagine a monster with three parts: head, torso and feet. Student A draws the head, then folds the paper over before passing it to student B, so student B cannot see what has been drawn. Then student B draws the torso and folds it again before passing it on to student C to add the feet.

Activity 2: Creating new monsters (with two students) Cut the monster into three sections and think how many new monsters (each with one head, one torso and one set of feet) can be created from a pair of monsters (See Figure 5.1).

Figure 5.1 Activity for creating new monsters

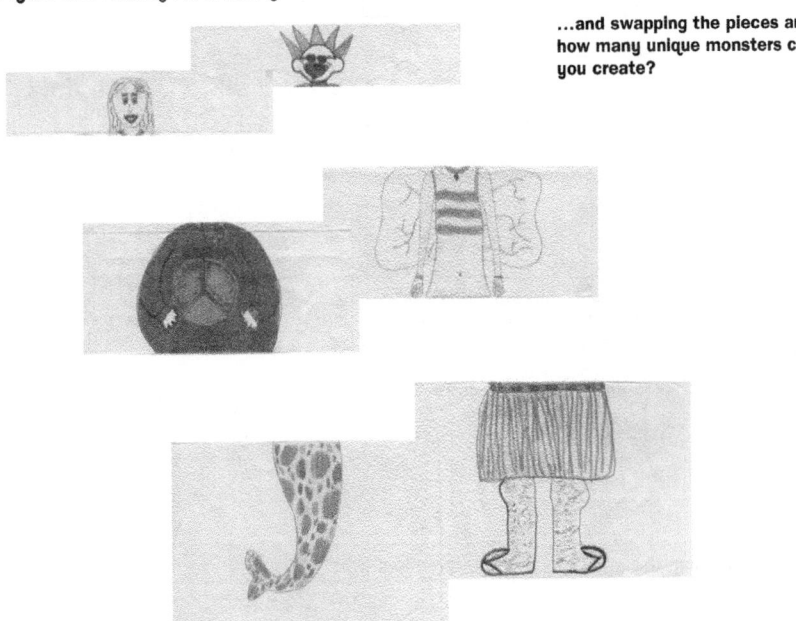

...and swapping the pieces around, how many unique monsters can you create?

Activity 3: Explaining answers – Reflections The activity is set up verbally with an emphasis on the process and on justification of how students arrive at their answers, either by manipulating the parts using a trial-and-error strategy or by guessing, with peer collaboration.

Modelling mathematically – a systematic method Then, the process of the teacher modelling in a mathematical way contains two steps: (1) modelling with a tree diagram to show the process pictorially (see Figure 5.2), relying on intuitive understanding, and (2) with an arithmetical explanation of the answer (see Figure 5.3): $8 = 2 \times 2 \times 2 = 2^3$.

Figure 5.2 Modelling the process using a tree diagram

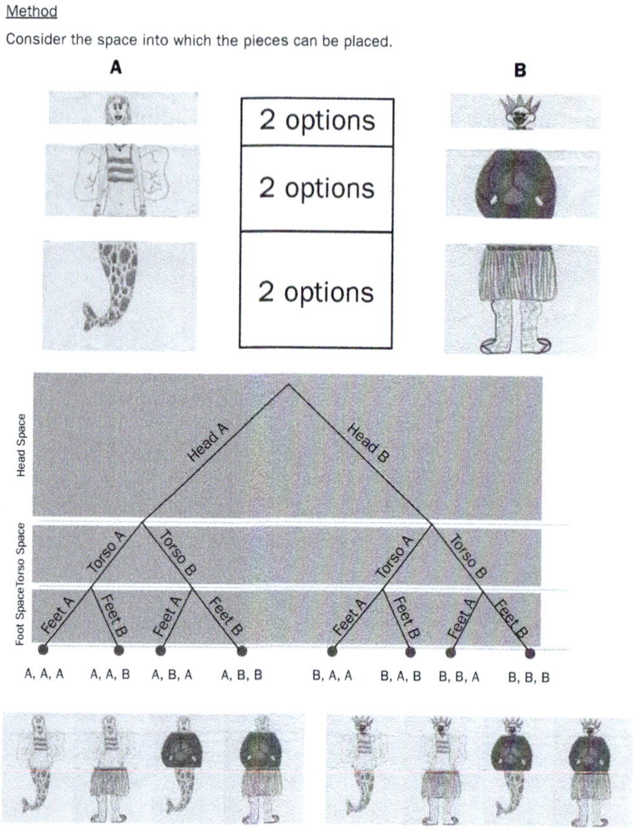

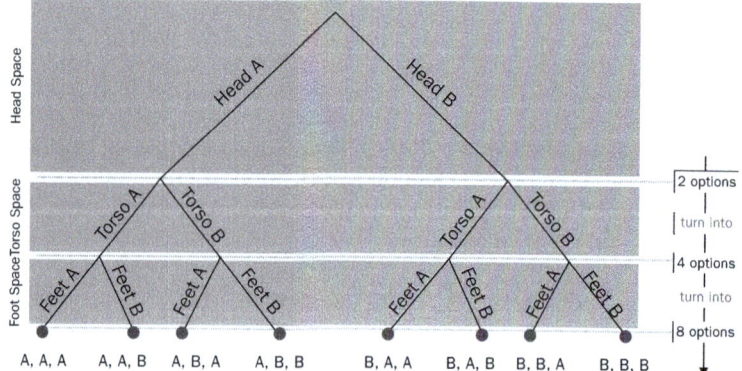

Figure 5.3 Explaining the answer 8 in an arithmetical way

Activity 4: Explaining the tree diagram – Reflections (see Figure 5.4)

Figure 5.4 Tree diagram reflection questions

What is the connection between this diagram and the task you've just done?

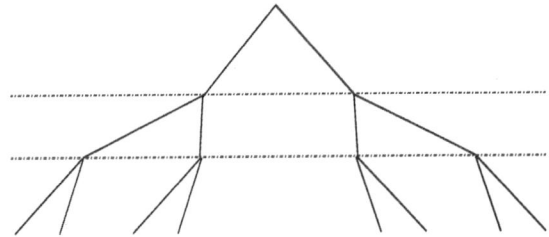

How could this diagram relate to a similar task?

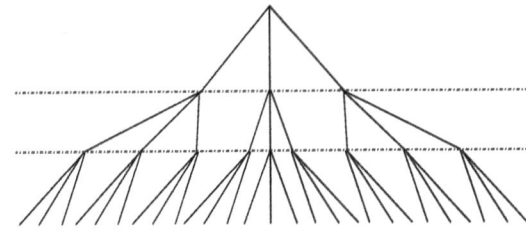

How many different creatures could be formed using 3 original monsters?

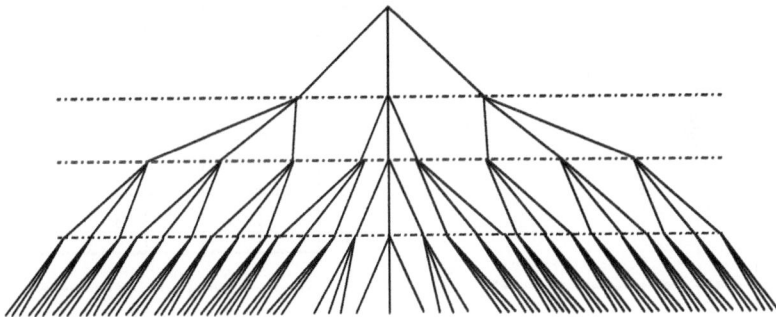

Activity 5: Generalizing the findings (see Figure 5.5) In summary, we have used reflection through a group of activities to reach generalization by explanation. Now the same pattern occurs in the Dice Series.

Figure 5.5 Generalizing the findings

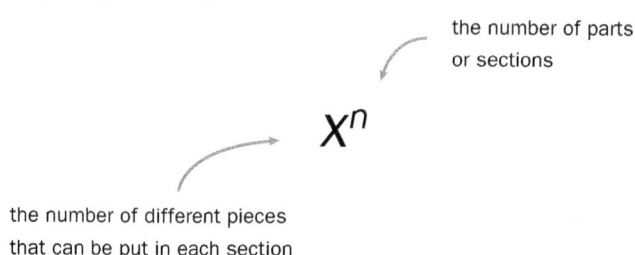

5.3.2 Dice series

Activity 1: Throwing dice Throwing a pair of dice and adding the two numbers together (see Figure 5.6) – how many outcomes are possible?

Figure 5.6 Dice

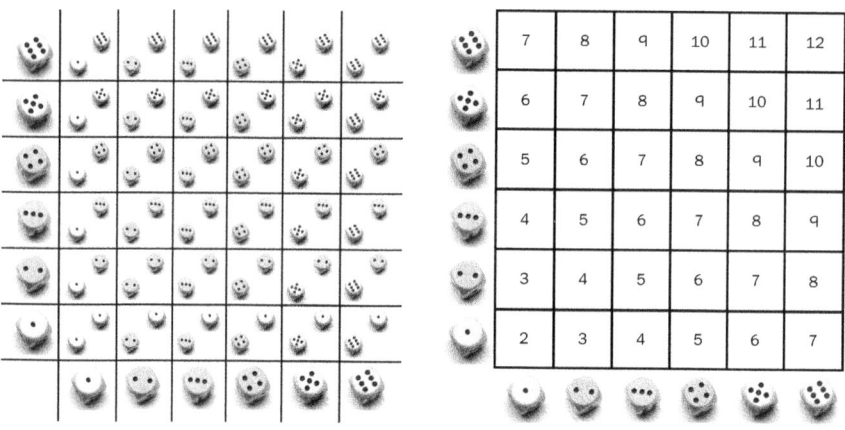

Activity 2: Explaining answers in a systematic way An example of a teacher's approach to modelling, using the Cartesian coordinate system.

Activity 3: Finding the most likely outcome and the least likely outcome

5.4 What comes next?

This chapter focuses on a lesson planning example employing the perspective of a visual conceptual structure. We use the CCF to link activities through a mastery approach, especially harnessing procedural variation (see section 3.2

for details), to lead students towards abstract understanding of relevant mathematical concepts. The dynamic core of the activities is a process, not a correct answer, and the planning is defined by the principles of generalization by explanation (see section 1.2) and causal connectivity (see section 4.2.4), rather than in terms of obtaining a certain piece of mathematical knowledge. The next chapter, however, actually details lesson planning with a particular piece of mathematical knowledge in mind, basic trigonometry. We will illustrate how the CCF helps with planning the hypothetical learning journey.

Practice activities

Creating concept diagrams for probability was part of chapter 4's practice activity. Figure 5.7 reveals one possible narrative for how this topic is built up.

Figure 5.7 Concept diagram

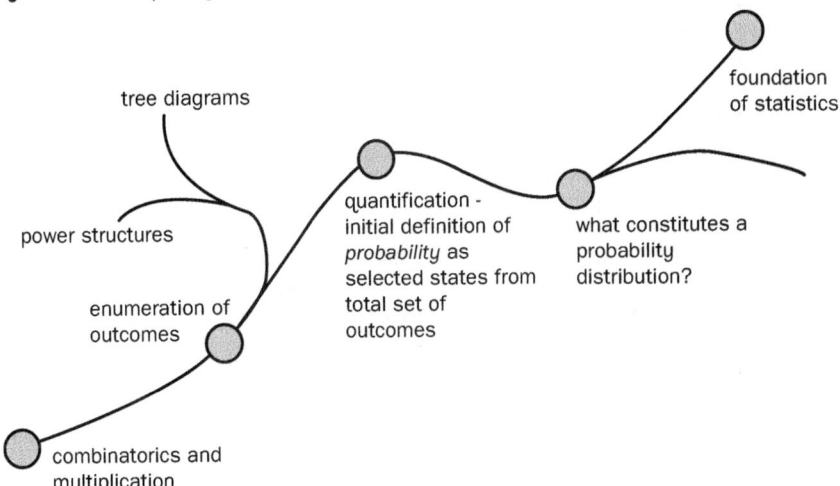

Reflect on the concept diagram you created for chapter 4 and compare it to the diagram above. List two points of similarity and/or difference.

1

2

References

Batanero, C. and Sanchez, E. (2013) What is the nature of high school students' conceptions and misconceptions about probability?, in G.A. Jones (ed.) *Exploring probability in school: challenges for teaching and learning*, pp. 260–89. Springer New York, NY: Kluwer Academic Publishers.

Cellucci, C. (2013) Top-down and bottom-up philosophy of mathematics, *Foundation of Science*, 18(1): 93–106. https://doi.org/10.1007/s10699-012-9287-6

Darragh, L. (2013) Constructing confidence and identities of belonging in mathematics at the transition to secondary school, *Research in Mathematics Education*, 15(3): 215–29. https://doi.org/10.1080/14794802.2013.803775

Dawson, J. and Wang, Y. (2019) How does a mastery lesson unfold? *Mathematics Teaching*, MT267: 11–4. Available at: https://durham-repository.worktribe.com/output/1290845 (accessed 6 September 2023).

Deieso, D. and Fraser, B. (2019) Learning environment, attitudes and anxiety across the transition from primary to secondary school mathematics, *Learning Environments Research*, 22: 133–52. https://doi.org/10.1007/s10984-018-9261-5

Ernest, P. (2018) The philosophy of Mathematics education: an overview, in P. Ernest (ed.) *The philosophy of Mathematics education today*, pp. 13–35. Cham, Switzerland: Springer.

Fischbein, E. (1977) Image and concept in learning Mathematics, *Educational Studies in Mathematics*, 8(2): 153–65. Available at: https://doi.org/10.1186/s40594-022-00328-0 (accessed 6 September 2023).

Giacomone, B., Beltran-Pellicer, P. and Godino, J. (2019) Cognitive analysis on prospective mathematics teachers' reasoning using area and tree diagrams, *International Journal of Innovation in Science and Mathematics Education*, 27(2): 18–32. Available at: http://funes.uniandes.edu.co/12997/ (accessed 6 September 2023).

Hiebert, J. and Lefevre, P. (1986) Conceptual and procedural knowledge in mathematics: An introductory analysis, in J. Hiebert (ed.) *Conceptual and Procedural Knowledge: The Case of Mathematics*, pp. 1–28. New York: Routledge.

Higgins, H., Martell, T., Waugh, D. et al. (2022) *Improving Mathematics in Key Stages 2 and 3*. EEF. Available at: https://d2tic4wvo1iusb.cloudfront.net/eef-guidance-reports/maths-ks-2-3/EEF-Improving-Mathematics-in-Key-Stages-2-and-3-2022-Update.pdf?v=1673008041 (accessed 5 June 2023).

Kaur, T., McLoughlin, E. and Grimes, P. (2022) Mathematics and science across the transition from primary to secondary school: a systematic literature review, *International Journal of STEM Education*, 9(1): 1–23. https://doi.org/10.1186/s40594-022-00328-0

McGalliard, M. and Wilson, P. (2016) Examining aspects of elementary grades pre-service teachers' mathematical reasoning, *Investigations in Mathematics Learning*, 9(4): 187–201. https://doi.org/10.1080/19477503.2016.1258857

Nunes, T., Bryant, P., Evans, D., Gottardis, L. and Terlektsi, M. (2014) The cognitive demands of understanding the sample space, *ZDM Mathematics Education*, 46: 437–48. https://doi.org/10.1007/s11858-014-0581-3

Ofsted (2021) *Research review series: Mathematics*. Available at: https://www.gov.uk/government/publications/research-review-series-mathematics/research-review-series-mathematics#curriculum-sequencing-conditional-knowledge (accessed 2 June 2023).

Rittle-Johnson, B., Schneider, M. and Star, J.R. (2015) Not a one-way street: Bidirectional relations between procedural and conceptual knowledge of mathematics, *Educational Psychology Review*, 27(4): 587–97. https://doi.org/10.1007/s10648-015-9302-x

Scottish Government (2015) *Play strategy for Scotland: Learning about play – investigating play through relevant qualifications in Scotland*. Available at: https://www.gov.scot/publications/play-strategy-scotland-learning-play-investigating-play-through-relevant-qualifications-scotland/pages/8/ (accessed 2 June 2023).

Tytler, R., Osborne, J., Williams, G. et al. (2008) Opening up pathways: Engagement in STEM across the primary-secondary school transition. *Canberra: Australian Department of Education, Employment and Workplace Relations*. Available at: http://hdl.voced.edu.au/10707/215301 (accessed 28 June 2023).

6 Basic trigonometry

Key arguments

This chapter focuses on an example of using the CCF with a more abstract concept, trigonometry. In contrast to chapter 5, this example reveals a top-down approach to exploring mathematics. We argue that although learning abstract concepts is typically thought of as part of developing students' procedural understanding, contextualization of these abstract concepts, for example through introducing the history of angles, can lead eventually to the *rethinking* of procedural understanding. In the case of trigonometry, there were historically two definitions: (1) trigonometry as a line system for a unit circle; and (2) trigonometry as a ratio system for two sides in a triangle. These two definitions have led to two kinds of conceptual understanding developing around the topic.

Key terms

- Procept

  Sine, cosine and tangent, on the one hand, are symbols for basic trigonometry. On the other, they point to the process for working out the relationship between two sides of a right-angled triangle and a relevant angle. The symbol and process are presented at the same time, and this phenomenon is called a *procept* by Gray and Tall (1994). Another example of a *procept* is a fraction such as $\frac{22}{7}$, which indicates a process of division as well as the answer of this division.

- Reflection on sorting and ordering

  When a piece of mathematical content is abstract in nature, a top-down approach to exploration aims to find key aspects within the concept. This process of searching is about prioritizing key information to work on. But behind this is a question of *why* to prioritize this particular information above other information. This is the sorting and ordering phase of the CCF, a part of the process based on students assessing criteria and showing reasoning. If the analysis and synthesis phase is about understanding *what* something is, then the sorting and ordering phase is about *why* it works in this particular way, or *how* it works. And if reflection in the analysis and synthesis phase means to generalize, then reflection in sorting and ordering is a process of deduction.

Basic trigonometry is one of the topics explored in the PANDA project. In 2015, there came into force a new maths national curriculum reform which introduced new topics (DfE 2021) which lay the ground for the project. After a discussion with leading practitioners (the heads of maths in local schools), local maths support teams (the Durham local authority and the Advanced Maths Support Programme) and maths educators at Durham University, three types of topics emerged as of particular interest: (1) hard-to-learn topics such as trigonometry and quadratic functions, (2) hard-to-teach ones such as vectors, and (3) ones that teachers think are neither hard to learn nor hard to teach, but that in reality students struggle with, such as linear graphs.

In this chapter, we focus on trigonometry, as it is viewed generally as a topic very few students liked and succeeded with. We first examine some of the topic's learning difficulties as identified in research, because these learning difficulties are associated with how we introduce the topic. We then look at trigonometry itself: what it is, and how we normally teach it. Finally, we introduce a sequence of four lessons used in the PANDA project aimed at reaching conceptual understanding, and we explain it with reference to the CCF.

6.1 Why trigonometry is hard to learn

The maths research community has long been pondering upon learning from the perspective of maths knowledge. Sfard (1991), for instance, distinguished two types of maths, operational and structural; and Gray and Tall (1994) proposed the notion of *procept*, a maths symbol which covers both process and concept. The notions of basic trigonometry, sine, cosine and tangent, are examples of the *procept*. They are each symbols, but also concepts and processes (the process of, for example, calculating sin30°). There are two approaches to defining trigonometry: (1) a ratio of two sides in a right-angled triangle, and (2) via a unit circle in the Cartesian coordinate system; for example, sine and cosine could be presented as coordinators, as well as horizontal and vertical distances (see 6.2.3 for detail). Either way involves an understanding of trigonometry's connections to its foundational topics, including ratio and angle (Brown 2005), and an ability to think of operations and object together. Gur (2009) argued that students needed to understand *why* this learning matters.

The question of 'why' is a central element of understanding. During a lesson observation of a Year 9 top set (the most able group of students) in north-east England, a student suddenly asked what the word 'trigonometry' means, and why we have it in mathematics. It was an unexpected question for the teacher, but exactly something that teachers should be prepared for.

6.2 What is trigonometry?

The term trigonometry is derived from the Greek words τρίγωνον (*trigōnon*, i.e. 'triangle') and μέτρον (*metron*, i.e. 'measure') (Barnard 2022). The invention of trigonometry goes back to Ancient Egypt and the Mediterranean world.

By the time that Hipparchus had compiled a list of 1,008 fixed stars, and Archimedes had figured out how to calculate π (Hogben 1968), astronomy had developed and there were increasing practical demands for world maps that could be used in navigation and land surveying. Both contributions provided the possibility of measuring the distance between two places on the earth. A key question at this point was how to measure the distance of the moon or the sun from the earth.

Measuring these heavenly distances began with the trigonometrical dictionaries that were compiled – the tables of sines, cosines, tangents, etc. These tables allowed for the solution of any arbitrary right-angled triangle, using a given ratio of sides to find all the angles, or vice versa (Duke 2011). Aristarchus made the first attempt to estimate the distance of the moon and the sun from the earth via cosine. Although the answer at that time was not particularly accurate, establishing the method marked a great achievement of Greek geometry and the beginnings of the language of trigonometry. But if this shows how trigonometry was born, how was this concept incorporated into the mathematics web of knowledge?

6.3 Teaching trigonometry

If the teaching of trigonometry can be said to have started with a unit circle whose radius is 1, then in this way, trigonometry was defined as a line of segment, a *line system*. Figure 6.1 shows a sine line and a cosine line, and Figure 6.2 shows a tangent line.

Figure 6.1 Sine line and cosine line in a unit circle

Figure 6.2 Tangent line in a unit circle

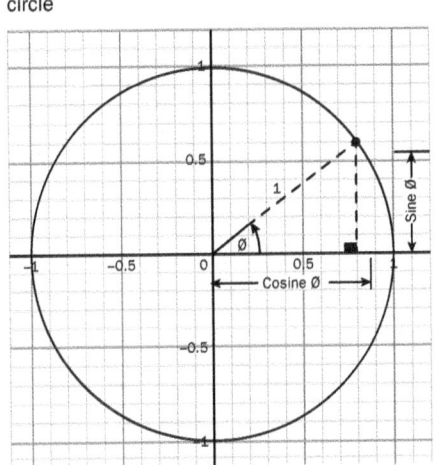

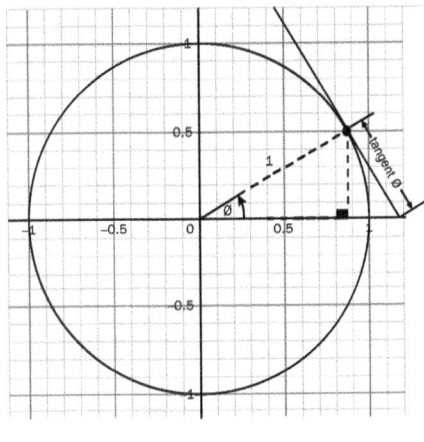

This definition of trigonometry takes a geometrical approach, focusing on the length of sides. Since the mid-eighteenth century, however, the definition shifted towards the algebraic rather than the geometric in European thought (Van Sickle 2011), with the definition later adapted to become a *ratio system*. The focus turned towards the ratio of two sides in defining an angle. To deepen students' procedural understanding of these ratios, formula triangles have been widely used in secondary schools.

Formula triangles are often used as a method to remember the proportional relationship between three variables, such as speed, distance and time: distance = speed × time. The method is used to find out the missing variable in a formula. Students are encouraged to use a finger to cover up the unknown letter, and the re-arranged formula appears (see Figure 6.3).

Figure 6.3 Using formula triangles

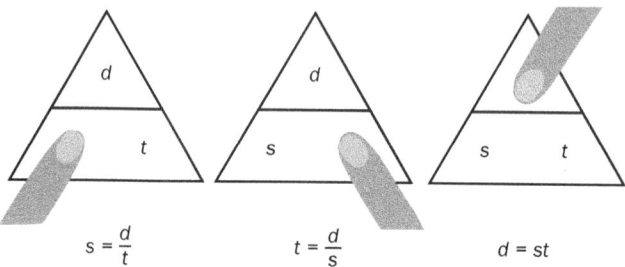

$$s = \frac{d}{t} \qquad\qquad t = \frac{d}{s} \qquad\qquad d = st$$

(Foster, 2021, p.30)

The same logic applies when operating with the three basic building blocks of trigonometry: sine, cosine and tangent (see Figure 6.4). In the example of sine, the first formula triangle in Figure 6.4 shows that by covering H (hypotenuse), students can immediately see that hypotenuse $= \dfrac{O \text{ (opposite)}}{S \text{ (sine)}}$. Foster (2021) criticized this method for the lack of emphasis placed on understanding. But what alternative ways could be employed?

Figure 6.4 Formula triangles for trigonometry

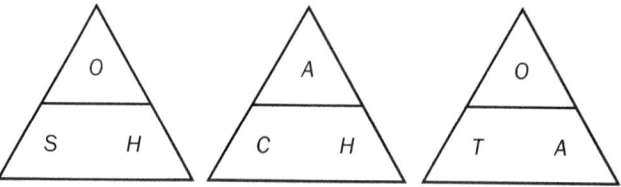

(Foster, 2021, p. 31)

6.4 Lessons in PANDA

Most schemes of work introduce Pythagoras' theorem before trigonometry. The same is true for the topic of similar triangles, which is related to scale factors. Whether these topics are arranged just before trigonometry or not, they can appear in the Relevance phase, in the first lesson, to check if students can label the three sides of the right-angled triangle properly. The sequence of the lesson starts from tangents, and tangents only. The plan builds up from the relevance phase towards the fourth phase, sorting and ordering (see Table 6.1). The same planning pattern appears in the second lesson, this time for sine and cosine (see Table 6.2). Again, at the delivery stage, these lessons may take longer than anticipated, and this could require adjustment locally.

Table 6.1 Outline of Lesson 1 – tanθ

Lesson 1 – tanθ	Phase of the Causal Connectivity Framework
Part 1: Pre-assessment Introducing tangent with similar triangles, from which the ratio system meaning of trigonometry is derived.	Relevance Highlighting cognitive (over affective) aspects.
Part 2: Finding missing angles Using a calculator to find an angle, checking the right mode (degree) to construct the concept image.	Analysis and synthesis Modelling the process and synthesizing different exploratory paths for the exercise question.
Part 3: Finding missing lengths There are three types, with missing lengths on the opposite and adjacent sides (with straightforward one-step questions and more complex two-step questions).	Sorting and ordering Giving students chance to articulate the rules they used to find the missing lengths, and to identify and critique the ratio structure of trigonometry.

Table 6.2 Outline of Lesson 2 – sinθ and cosθ

Lesson 2 – sinθ and cosθ	Phase of the Causal Connectivity Framework
Part 1: Pre-assessment Introducing sine and cosine through pairs of similar triangles.	Relevance
Part 2: Finding missing angles Group activity involving the unit circle in the Cartesian coordinates system, which links with the next lesson.	Analysis and synthesis
Part 3: Finding missing lengths 'Have a think' activity, which links with the next lesson.	Sorting and ordering

The third planned lesson (see Table 6.3) adds a unit-circle approach to support the causal connectivity that is to come later. The focus is on the exact value of sinθ and cosθ for θ = 0°, 30°, 45°, 60°, 90°, and tanθ for θ = 0°, 30°, 45°, 60°. Rather than asking students to memorize these values, the lesson design focuses on *how*, asking students to reason out these values using Pythagoras' theorem. For 30° and 60°, it adds another kind of proof, using unit circle, to set up the causal connectivity phase.

Table 6.3 Outline of Lesson 3 – Exact value

Lesson 3 – Exact value	Phase of the Causal Connectivity Framework
Part 1: Trigonometry in unit circle Sine line, cosine line and tangent line, from which the line system meaning of trigonometry is derived.	Relevance Highlighting cognitive (over affective) aspects
Part 2: The exact values of sinθ and cosθ for θ = 0°, 90°, and tanθ for θ = 0°	Analysis and synthesis Using different types of representation to develop structural generalization.
Part 3: The exact values of sinθ, cosθ and tanθ • for θ = 45° • for θ = 30°, 60° (using the unit circle method)	Sorting and ordering Using two different methods to develop students' ability to understand other rules and other forms.

The final lesson (see Table 6.4), problem-solving, selects activities that look towards the final two phases, Causal connectivity and Proof or theorizing, with the aim of leading students to integrate the new concepts of trigonometry into their web of mathematical knowledge. The formula triangle method is now introduced, so here at the end of the sequence, students are provided with procedural understanding, but only after conceptual understanding has been properly built up.

Table 6.4 Outline of Lesson 4 – Problem-solving

Lesson 4 – Problem-solving	Phase of the Causal Connectivity Framework
Part 1: Pre-assessment Introducing formula triangles Labelling the sides in a right-angled triangle (opp, adj, hyp), recalling trigonometric ratios, and using formula triangles	Causal connectivity After establishing an understanding of the line system and ratio system meanings of trigonometry, showing a connection between *conceptual* understanding and *procedural* understanding by using formula triangles.
Part 2: Links with isosceles triangles	Proof or theorizing Decontextualizing in-context mathematical tasks with the link of properties of isosceles triangles.

6.4.1 Lesson 1 – Tangent

Part 1: Pre-assessment
Example 1: $\triangle ABC$ and $\triangle ADE$ are similar triangles. $BC = 1.5$ cm, $DE = 9$ cm, $AB = 2$ cm.
Calculate the length of AD (see Diagram 6.1).

Diagram 6.1

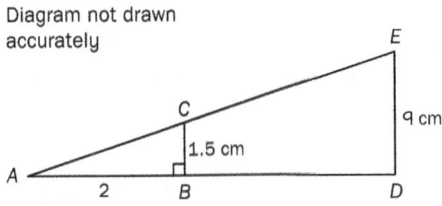

Diagram not drawn accurately

Conclusion: The ratio 'opposite: adjacent' of Angle A is consistent.

<u>Example 2</u>: Find the length of AE and AG, if the ratio 'opposite:adjacent' of angle θ is consistent (see Diagram 6.2).

Diagram 6.2

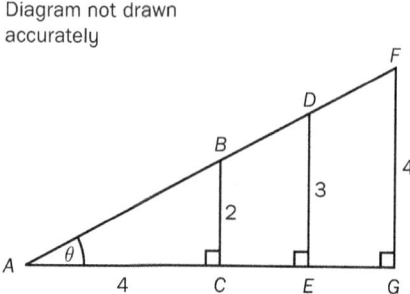

Diagram not drawn accurately

Conclusion: This ratio is a trigonometric ratio called the tangent ratio, tanθ = opposite ÷ adjacent

<u>Question</u>: Can you find the value of tan$A$ from Example 1, and tanθ from Example 2?
Teacher to explain how to use a calculator to find the size of angle A, and ask students to find the size of angle θ.

Part 2: Find the missing angles
<u>Exercise 1</u>: Find the size of the missing angle at B (see Diagram 6.3).

Diagram 6.3

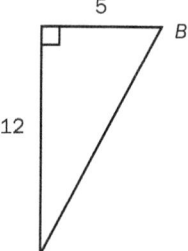

<u>Exercise 2</u>: $ABCD$ is a rectangle where $BC = 3AB$. P and Q are points on BC such that $BP = PQ = QC$ (see Diagram 6.4). Show that $\angle DBC + \angle DPC = \angle DQC$

Diagram 6.4

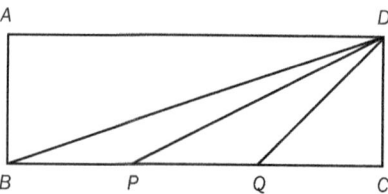

Part 3: Find the missing lengths

3.1 Missing lengths on the opposite side
<u>Example 1</u>: Use the tangent ratio to calculate the length labelled x in this right-angled triangle (see Diagram 6.5).

Diagram 6.5

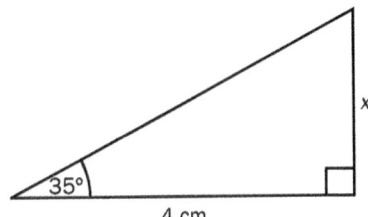

4 cm

<u>Exercise 1</u>: Calculate length x in this a right-angled triangle (see Diagram 6.6).

Diagram 6.6

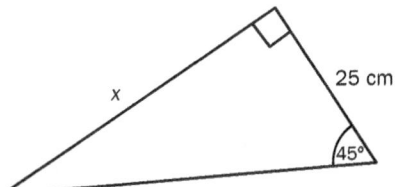

<u>Exercise 2</u>: This shape is made from two right-angled triangles (see Diagram 6.7).

1   Calculate the length labelled a. Give your answer to 1 decimal place.
2   Calculate the length labelled b.

Diagram 6.7

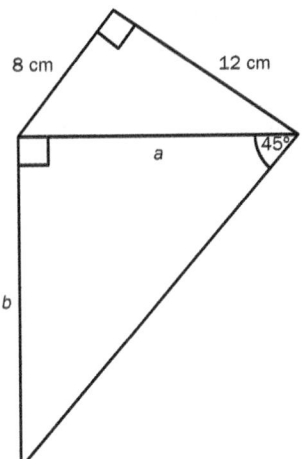

3.2 Missing lengths on the adjacent side
<u>Example 2</u>: $\triangle ABC$ is a right-angled triangle.
$AB = 5.4$ cm and $\angle ACB = 62°$ (see Diagram 6.8).
Calculate the length of BC. Give your answer correct to 2 significant figures.

Diagram 6.8

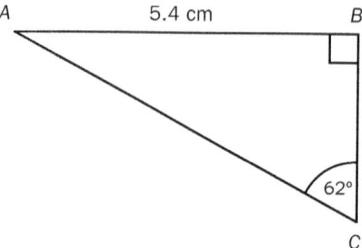

<u>Exercise 3</u>: $\triangle ABC$ is a triangle, with line BD perpendicular to AC (see Diagram 6.9).
$\angle BAC = 43°$, $BD = 8$ cm and $AC = 12$ cm. Calculate $\angle BCA$.

Diagram 6.9

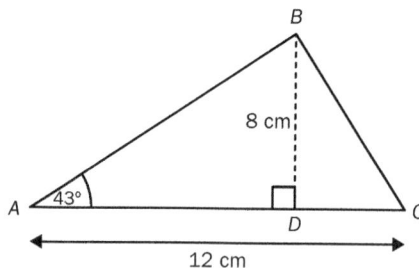

6.4.2 Lesson 2 – Sine and cosine

Part 1: Pre-assessment Look at the following two sets of triangles (see Diagram 6.10). Could you prove that they are similar triangles? What other conclusions can you prove?

Diagram 6.10

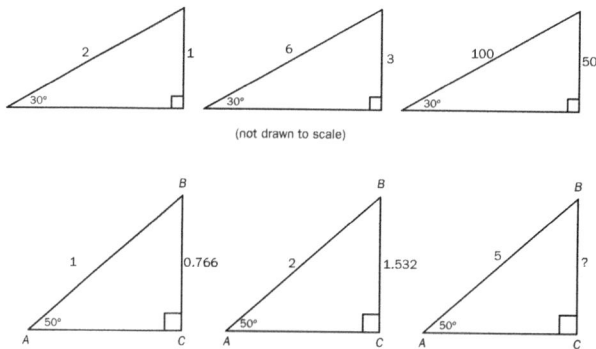

The sine ratio: $\sin\theta = \dfrac{\text{opposite}}{\text{hypotenuse}}$

The cosine ratio: $\cos\theta = \dfrac{\text{adjacent}}{\text{hypotenuse}}$

Part 2 – Find the missing angles

<u>Example 1</u>: Find the size of missing angle x (see Diagram 6.11).

Diagram 6.11

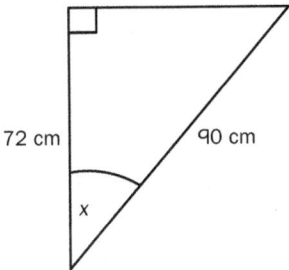

Now move x so that 72 cm is the opposite side. Can you find the size of angle x?

<u>Exercise 1</u>: Find the size of missing angle x (see Diagram 6.12).

Diagram 6.12

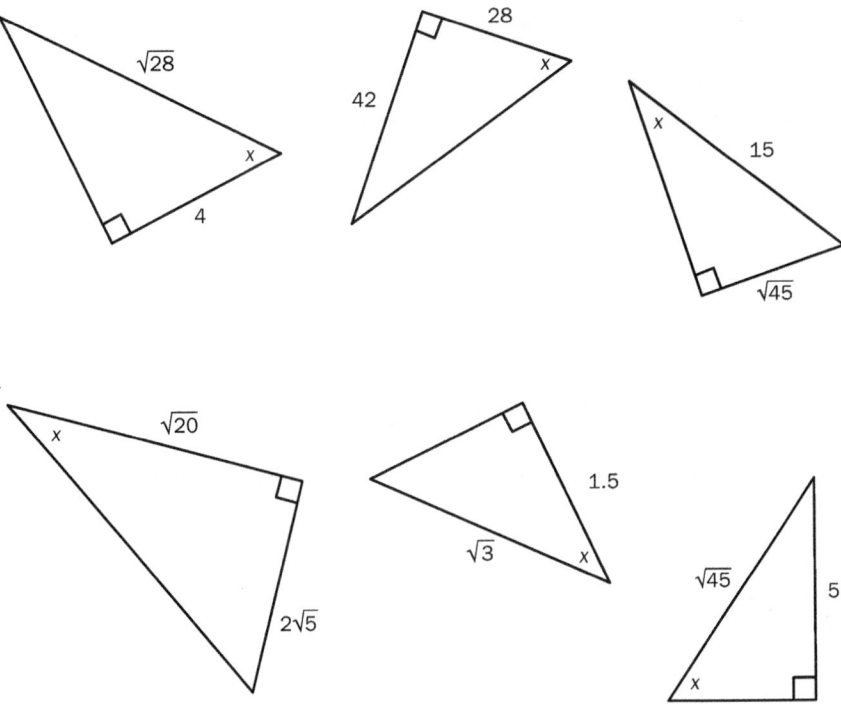

<u>Group activity</u>: Find as many missing angles as possible from the diagram (see Diagram 6.13).

Diagram 6.13

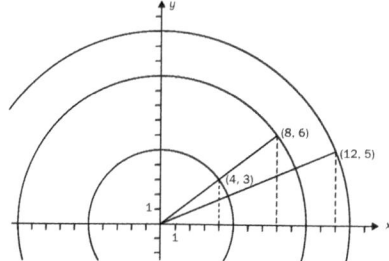

Part 3: Find the missing lengths

<u>Example 1</u>: Find the missing length x (see Diagram 6.14).

Diagram 6.14

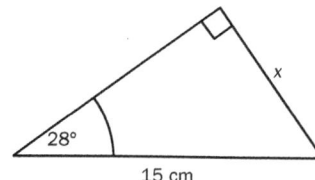

<u>Consider</u>: What if the missing length were located on another right-angled side? Can you still find the missing length?

<u>Exercise 1</u>: Mary says that you can calculate x using the cosine ratio, but Sarah says that you can use the sine ratio (see Diagram 6.15). Explain why both of them are correct.

Diagram 6.15

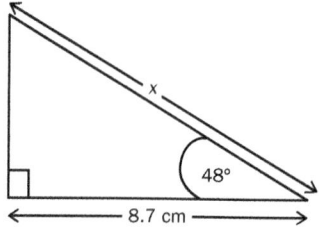

<u>Consider</u>: What is the relationship between sin 48° and cos 42°? Can you generalize it?

sin 1° = cos _____ °
sin 30° = cos _____ °
sin 60° = cos _____ °
sin 0° = cos _____ °
sin 90° = cos _____ °
sin 45° = cos _____ °

6.4.3 Lesson 3 – Exact values

Part 1: Trigonometry in Unit Circle The radius of the circle is 1 unit long (see Diagram 6.16).

Diagram 6.16

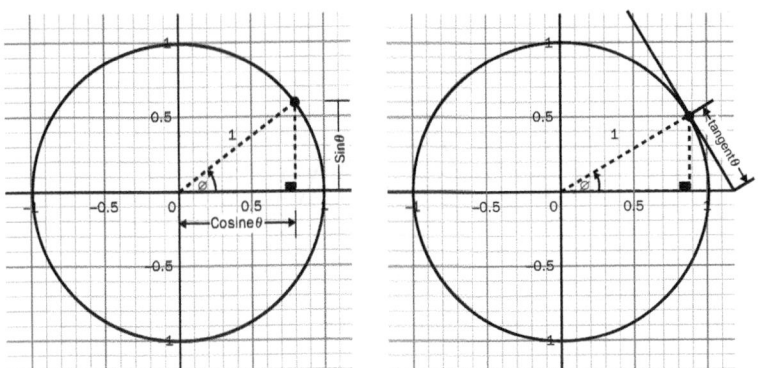

Consider: Why is a line that touches a circle at a point/is perpendicular to the radius through this point called a tangent line?

Part 2: Exact value of sin θ and cos θ for $\theta = 0°$, 90°, tan θ for $\theta = 0°$ Consider this unit circle.
When $\theta = 0°$, what happens with the sine line and the cosine line?
sin 0° = 0 and cos 0° = 1
When $\theta = 90°$, what happens with the sine line and the cosine line?
sin 90° = 1 and cos 90° = 0
When $\theta = 0°$, what happens with the tangent line?
tan 0° = 0

Part 3 – The exact values of sin θ, cos θ, and tan θ for $\theta = 30°$, 45° and 60°
- For $\theta = 45°$

  Explore: Use the isosceles right-angle triangle (see Diagram 6.17) and Pythagoras' theorem to find the exact values when $\theta = 45°$.

Diagram 6.17

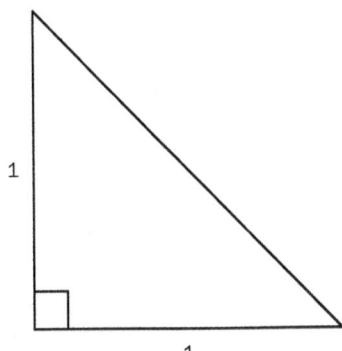

$$\sin 45° = \frac{1}{\sqrt{2}} = \frac{\sqrt{2}}{2}; \cos 45° = \frac{1}{\sqrt{2}} = \frac{\sqrt{2}}{2};$$

$$\tan 45° = \frac{1}{1} = 1$$

- For θ = 30°, 60°
  1 Traditional method – using the properties of equilateral triangles (see Diagram 6.18)

Diagram 6.18

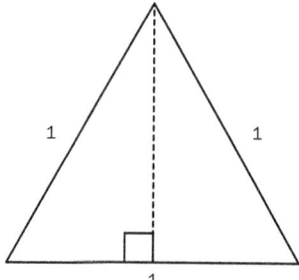

  2 Unit circle method – using circle theorem 4 (see section 2.3.4), ∠ACB = 90° (see Diagram 6.19)

Diagram 6.19

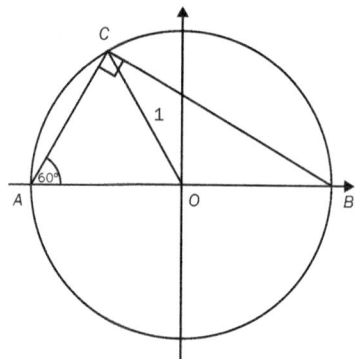

Conclusions

	θ = 0°	θ = 30°	θ = 45°	θ = 60°	θ = 90°
sin θ	0	$\frac{1}{2}$	$\frac{\sqrt{2}}{2}$	$\frac{\sqrt{3}}{2}$	1
cos θ	1	$\frac{\sqrt{3}}{2}$	$\frac{\sqrt{2}}{2}$	$\frac{1}{2}$	0
tan θ	0	$\frac{\sqrt{3}}{3}$	1	$\sqrt{3}$	

6.4.4 Lesson 4 – Problem-solving

Part 1: Formula triangle (see Diagram 6.20)

Diagram 6.20

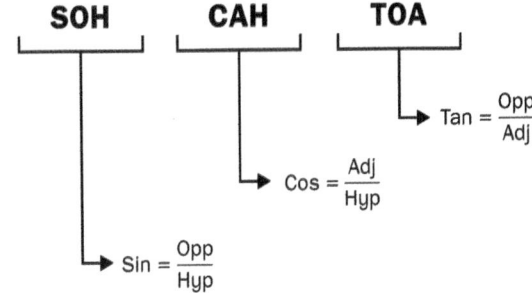

SOH   CAH   TOA

$$Tan = \frac{Opp}{Adj}$$

$$Cos = \frac{Adj}{Hyp}$$

$$Sin = \frac{Opp}{Hyp}$$

<u>Discussion</u>: how to find the length of *CD* in Diagram 6.21.

Diagram 6.21

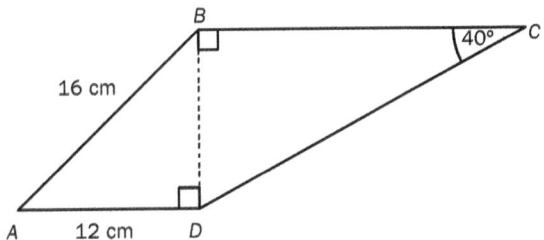

Part 2: Links with isosceles triangles
<u>Example 1</u>: What do you know and what can you work out?
Determine the length of the base of the triangle in Diagram 6.22.

Diagram 6.22

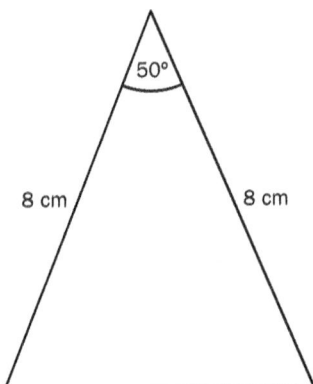

Further question 1:
Sarah makes a kite from two isosceles triangles (see Diagram 6.23). Find the height *AC* of the kite. Give your answer to the nearest centimetre.

Diagram 6.23

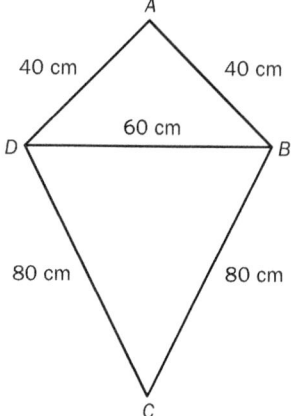

Further Question 2:
In Diagram 6.24, how could you calculate angle α and angle β, if the diagonal *BD* has length 10 cm? Give your answers correct to 1 decimal place.

Diagram 6.24

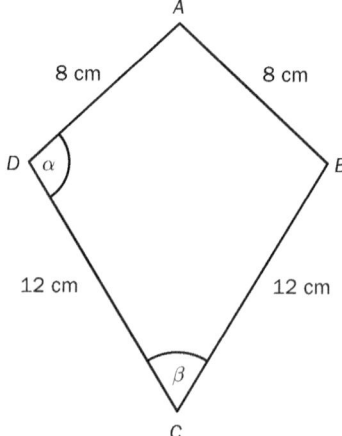

6.5 What comes next?

In this chapter, we have discussed the topic of trigonometry in a sequence of lessons designed using the CCF. Reasoning here emerges in two dimensions: from linking activities to expand student understanding through the coherence of the format of activities, and from the activities themselves promoting the drawing of logical links. At secondary level, teaching more abstract topics, such as trigonometry, requires particular concern for building upon students' conceptual understanding. To achieve this, we have modelled a possible way of planning lessons with a focus on reasoning. Of course, this is not all down to teachers. In the next chapter, we look more broadly at assessment around reasoning, and how school leaders can support this.

Practice activities

Returning to the concept diagram for basic trigonometry from the practice activity in chapter 4, Figures 6.5, 6.6, and 6.7 add three possible narratives concerning how this topic can be built up. Rank these diagrams in whichever way you see fit, and write down the criteria you used to assess them.

Figure 6.5 Narrative 1 for the topic of trigonometry

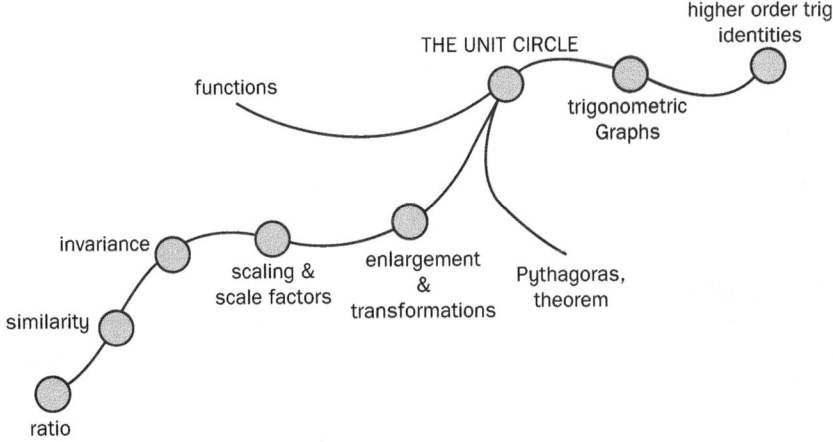

Figure 6.6 Narrative 2 for the topic of trigonometry

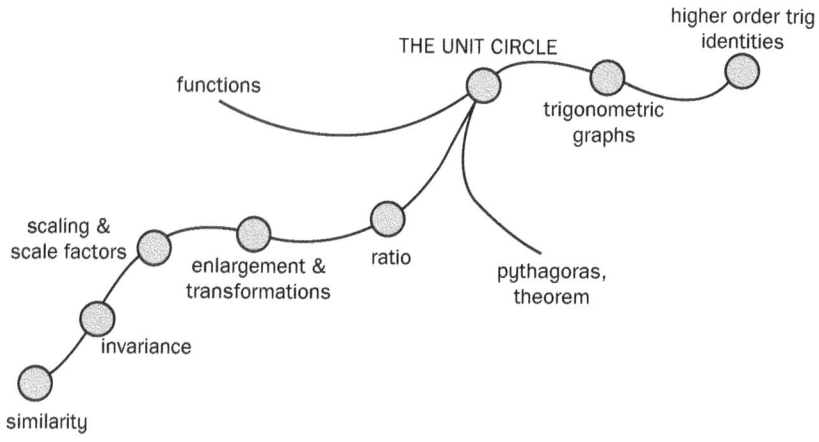

Figure 6.7 Narrative 3 for the topic of trigonometry

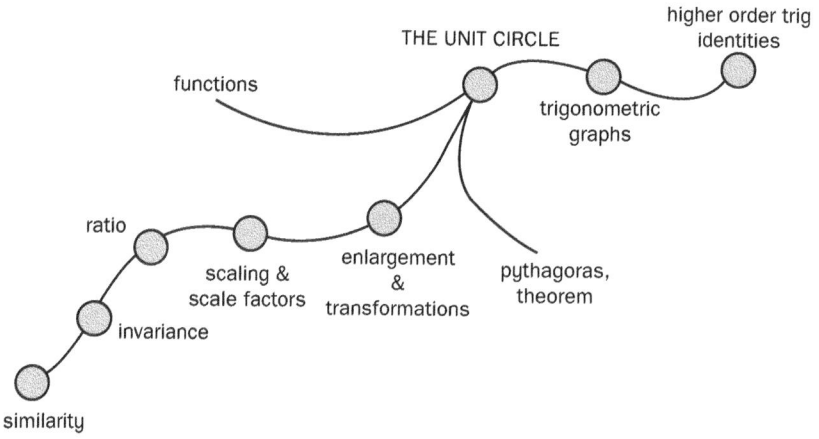

References

Barnard, R.W. (2022) *Trigonometry*. Available at: https://www.britannica.com/science/trigonometry (accessed 6 June 2023).

Brown, S.A. (2005) *The trigonometric connection: Students' understanding of sine and cosine.* (Publication No. 3233908.) Doctoral dissertation, Illinois State University. ProQuest Dissertations and Theses Global. Available at: https://www.proquest.com/docview/304986316/previewPDF/57805F44BB724F61PQ/1?accountid=14533 (accessed 6 September 2023).

Department for Education (2021) *Education reform: new national curriculum for schools.* Available at: https://www.gov.uk/government/speeches/education-reform-new-national-curriculum-for-schools#:~:text=From%20September%202015%2C%20the%20 new,the%20library%20of%20the%20House (accessed 6 June 2023).

Duke, D.W. (2011) The Very Early History of Trigonometry, *International Journal of Scientific History*, 17: 34–42. Available at: https://people.sc.fsu.edu/~dduke/early-trig12.pdf (accessed 5 June 2023).

Foster, C. (2021) *On hating formula triangles*. Loughborough University, journal contribution. Available at: https://hdl.handle.net/2134/13574498.v1 (accessed 06 June 2023)

Gray, E.M. and Tall, D.O. (1994) Duality, Ambiguity, and Flexibility: A Proceptual View of Elementary Arithmetic, *Journal for Research in Mathematics Education*, 26(2): 114–41. https://doi.org/10.5951/jresematheduc.25.2.0116

Gur, H. (2009) Trigonometry Learning, *New Horizons in Education*, 57(1): 67–80. Available at: https://files.eric.ed.gov/fulltext/EJ860819.pdf (accessed 3 June 2023).

Hogben, L.T. (1968) *Mathematics for the Million*. New York: WW Norton & Company.

Sfard, A. (1991) On the dual nature of mathematical conceptions: Reflections on processes and objects as different sides of the same coin, *Educational Studies in Mathematics*, 22: 1–36. https://doi.org/10.1007/BF00302715

Van Sickle, J. (2011) *A History of Trigonometry Education in the United States: 1776–1900*. Columbia University. Available at: https://www.proquest.com/docview/867674316?pq-origsite=gscholar&fromopenview=true (accessed 5 June 2023).

Part 3

In Between Theory and Practice

7 Assessing reasoning

Key arguments

This chapter offers a look at how to student reasoning is assessed. Usually, assessment of reasoning emphasizes a particular sort of answers, such as proof (in topics like congruent triangles). This shows the power of assessing. But we argue that this can become a prison, too, when a particular kind of answer is seen as the be-all and end-all of teaching and learning mathematics. We suggest that reasoning can become even more powerful when employing methods like scoring rubrics to describe and assess reasoning as a *process*. The CCF brings the design of reasoning experiences into the assessment picture; with this, assessing no longer has to be something *outside* of teaching and learning.

Key term

- Scoring rubrics for mathematics reasoning
  In scoring rubrics, mathematics reasoning is divided into component parts such as analysing, generalizing, and judging. Each of these parts of reasoning may be assessed according to levels from *extending* to *not evident*. To assess reasoning, both the structure of the reasoning and how it progresses must be considered.

This chapter starts from considering how reasoning is assessed from the point of view of research methodology, mainly looking at intervention studies and comparative judgements. Neither is part of normal teaching practice or integrated into the learning journey, since they rely on third parties taking snapshots to check reasoning. But awareness of these methods for assessing reasoning in academic contexts can nonetheless help broaden teachers' understanding.

We then explore daily classroom assessment and, in particular, the practice of using scoring rubrics. This brings the checking of reasoning a step closer to the learning journey. The last section discusses precisely what to assess when we are interested in reasoning. But the idea of 'what to assess' actually follows a traditional view of assessment that conceives of it as separate from the learning journey. How, then, can we account for students' learning experiences while avoiding common traps around assessment? A key contribution of the CCF is to show how learning and assessment can be linked together better. We start with a short quiz, all of the answers to which can be found in the chapter.

Box 7.1 Always, sometimes or never true

1 Assessment is a check point to discover what students have not mastered yet.
2 Assessment of reasoning skills is closely related to mathematics topics.
3 In the daily classroom, I spend more time checking students' conceptual understanding and their basic mathematics skills than assessing their reasoning skills.

7.1 Assessing from a research point of view

In this section, the focus is not merely the two methodologies for assessment, but also two different views of assessment that lie behind them. There is a great difference between a structured measurement, as used in an intervention study, and the one that focuses on the quality of an answer or a process itself, as in a comparative judgement. Likewise, there are important differences between the clear targeting of pre-determined aims and less structured processes. And it is also necessary to distinguish between assessment that defines answers as right and wrong according to a marking scheme and one that views solutions from a holistic perspective in comparison with other possible solutions.

7.1.1 Assessing reasoning itself

First, we will introduce some findings from intervention projects, those that relate closely to improving students' reasoning abilities. The key idea is that addressing reasoning abilities in this way is often complicated, as this kind of project relies fundamentally on a marking scheme and leaves little flexibility over what and how to assess. We then introduce comparative judgements as an alternative approach to marking work, one which may also point to a possible way of reconsidering the fundamentals of assessing reasoning.

Intervention studies One way of assessing reasoning is to address the development of students' cognitive development, as it is understood from the perspective of two main educational theories, those developed by Piaget and by Vygotsky. While Piaget and his collaborators began to look at reasoning from the individual student's perspective (e.g. considering their problem-solving strategies and their justifications for using these strategies), Vygotsky provided socio-cultural tools concerning issues such as how to organize students to use different representations in expressing their thoughts and to start internalization during their learning journey (Brodie 2009). These two main educational theories provide empirical basis for judgments on (1) when to develop core aspects of certain type of reasoning, e.g. by the age of 11, students should be able to do formal operations, according to Piaget; and (2) how mathematical topics can be built up towards abstract levels of understanding via reasoning, e.g. using different representations.

The design of intervention studies divides a whole sample of students into two groups: the intervention group (we might call it the 'experiment' group) and the control group (or the 'business as usual' group), and the results each group achieves in pre- and post-intervention assessments are compared. This methodology falls into the category of experimental or quasi-experimental studies. If random allocation of the sample into two groups is possible, then the study is called a randomized controlled trial (RCT), a research design seen as a gold standard (albeit that this status is debatable) (Grossman and Mackenzie 2005). The core value of the assessment here is to ascertain if particular pre-determined targets are met, or the extent to which they are met and for which particular group of students.

Let's take as an example spatial reasoning, since we discussed various levels of reasoning in relation to this topic in chapter 1. Lowrie et al. (2018) suggest that a classroom-based intervention should be combined with a pedagogy framework, such as experience-language-pictorial-symbolic-application (ELPSA, which we discussed in chapter 2 when considering structure-oriented approaches). These researchers set up two studies, one for a primary setting (Lowrie et al. 2018) and the other for a secondary setting (Lowrie et al. 2019), aiming to assess the effectiveness of the intervention (in this case deployment of the ELPSA pedagogy framework) in developing students' spatial reasoning and improving their mathematics performance.

The assessment is called a spatial reasoning instrument (SRI). It is suitable for students aged 9 to 12 and contains three skills: mental rotation (including 2-D and 3-D rotation), spatial orientation (such as representing 3-D space on a 2-D plane), and spatial visualization (using origami). In this way, it exemplifies intervention-based research that targets certain reasoning areas. The programme engages teachers with a pedagogy framework and then measures if using it leads to real changes in their practice which ultimately leads to students' learning outcome. The results were that the intervention group showed a statistically significant improvement in spatial visualization and spatial orientation, but also that these developments in spatial reasoning were associated with improvement of mathematics performance with the content of geometry and measurement, and number and algebra (Lowrie et al. 2019). While this intervention proved useful in teaching spatial reasoning, the assessment part of the research in particular spliced reasoning with associated topics (see chapter 1 for discussion of different types of reasoning). We put this idea on hold for a moment while we introduce our other assessment method, comparative judgement.

Comparative judgement Nunes and Csapó (2011) argued that to design assessment in connection to certain types of reasoning, we should ask students to use different types of representations, such as drawing, and to apply the reasoning in different situations, such as real-life ones. The purpose is to minimize reliance on reading skills. This reflects a valuing of diverse answers rather than answers prescribed by standard marking schemes, and ultimately it recognizes the need for an alternative approach to assessment. To understand this, it is first necessary to have a clear sense of what *normal* practices of assessment are like. The most common type, normative assessment (for example, end-of-term assessments), focuses on students' levels of achievement of learning goals in comparison with their peers. In contrast, criterion-based assessment (for example, that using a scoring rubric), supports teachers in targeting specific areas (see section 7.2). However, both are similar in that they look towards structured knowledge and skills. But they do little to address how we might assess students' conceptual understanding, their problem-solving skills, their creativity and their reasoning, all of which are not structured in the same way. The method of comparative judgement explores some of these possibilities.

Comparative judgement involves holding up work from two students against each other, with a number of experts comparing it and producing a judgement as to which one is better according to criteria for the task. These decisions are processed into a statistical model, producing a parameter estimate to rank the order of student work and generate test results (Pollitt 2012). This method has been used to assess students' mathematics problem-solving, proving its reliability and validity (Jones and Inglis 2015). Another comparative judgement study (Bisson et al. 2016) assessed students' conceptual understanding of letters in algebra, using 10 mathematics PhD students as experts. The results also showed moderate correlation with the existing instrument and higher inter-rater reliability.

Let's take as an example proof, which we discussed in chapter 2. The capacity to establish proof starts with the process of structural generalization through making reasoned conjectures from empirical generalization. A crucial aspect of proof to be aware of, however, is precisely how students understand what they are doing when they set about proving. Research has found that student teachers in their first year of study in secondary mathematics education acknowledge the importance of proof, but that they are concerned that the proof taught in secondary schools mainly rests on imitative reasoning and that this causes difficulties when moving on to university study (Basturk 2010). Imitative reasoning (IR) includes memorized reasoning (MR) and algorithmic reasoning (AR), while creative reasoning (CR), as discussed in the first chapter, can be fostered by looking at non-routine problems.

So, what tends to be valued the most in terms of proof? Davies, Alcock and Jones (2021) used comparative judgement to identify the features collectively considered by expert as well as non-expert judges to be prevailing characteristics. Their study showed that argumentation, involving the logical structure of the proof, and conviction, such as using contradiction, is the root of mathematicians' view of proof. Moreover, Davies, Alcock and Jones (2020) suggested that comparative judgement could provide way of assessing the construction of proof as a product. So we have some hopes here.

7.1.2 A wider view of assessing

If assessment in the classroom supports individual students' learning and helps improve teaching practice, then large-scale assessment monitors education systems overall, something which also has an important impact on teaching and learning (Suurtamm et al. 2016). In between the two come school leaders, who provide a systematic vision for how different parts are integrated into their school environment, a topic for discussion in the next chapter. Assessing reasoning is closely linked to surrounding contextual factors, and findings from the OECD point to how reasoning takes shape globally.

Mathematical reasoning in PISA 2021 appears in four types of data collected:

- Data on student performance: how well students can reason based on their understanding of key school mathematics.
- Data about what is available to support students in reasoning: to what extent schools provide opportunities for students to learn to reason.
- Data concerning students' attitudes to reasoning tasks.
- Data on teachers' views about the role of reasoning in their teaching (OECD 2018).

So, from previous PISA assessment cycles, what can we tell about the sorts of environment that appear to support reasoning? How might teachers' beliefs influence their teaching with reasoning? On the classroom level, teachers seem to spend more time on understanding than on reasoning; roughly 55 per cent of students have reported being asked to present their thinking or reasoning, while 70 per cent of students reported being asked if they had understood what was taught (OECD 2016). If we take a step back to look at the environment (i.e. the school level), we find that academically oriented schools with less behaviour disruption are more likely to use cognitive-activation strategies, which focus on thinking and reasoning, as students here may not need too long to learn basic concepts (OECD 2016).

The OECD's Teaching and Learning International Survey (TALIS) seeks the perspectives of school teachers and leaders on teaching practices, the school environment, and so on. It asks for teachers' beliefs about the nature of teaching and learning, beliefs which are shaped by the cultural and pedagogical factors surrounding their education systems. The first cycle of TALIS, in 2008 (OECD 2009), placed beliefs about student learning and instruction into two categories: direct transmission and constructivist views. The direct transmission view sees teachers expected to communicate knowledge in 'a clear and structured way', while the constructivist view emphasizes the search for knowledge (OECD 2009, p. 92). When teachers hold constructivist beliefs about teaching, the processes of thinking and reasoning become more important than mastering specific curriculum content (OECD 2009). Teachers subscribing to direct transmission see curriculum knowledge as about presenting truth and fact, the key aim being to master specific knowledge and skills.

But if limited only to what is in the curriculum, how could previous generations have created this knowledge? The constructivist view holds up the process of acquiring knowledge as the purpose of learning. The next TALIS cycle, in 2013, reported that over 80 per cent of teachers agree that thinking and reasoning are more important than specific content knowledge in the curriculum (OECD 2014). This is not an either-or situation, but a step towards more fully developed thinking on assessment. It must be mentioned that there is also an institutional-level impact on pedagogical practice and professional characteristics (Ainley and Carstens 2019).

Global Teaching InSights (OECD 2020) took a unit on quadratic equations as the material for a comparative video study of teaching, the aim being to gauge learning opportunities for students and to offer insight into differences practices occurring in different places. Teachers in Chile, Spain and Shanghai (China) encouraged most reasoning, for example around determining the solution of a given equation.

In case of quadratic function, the teaching materials in the three territories provided most opportunities for reasoning about the relationship between solutions of equations and types of equation by calculating $b^2 - 4ac$ and comparing the result with 0 to determine whether there is one solution (i.e. two solutions with the same value), two solutions or no solutions.

The study looked at four learning opportunities, algebraic procedures, applications, functions (such as graphs of quadratic functions), and reasoning across eight territories: Chile, Colombia, England, Germany, Japan, Madrid (Spain), Mexico and Shanghai (China). Common practice across these places was to emphasize algebraic procedures. The proportion of teaching materials related to reasoning were highest in Shanghai and Madrid, while Japan, Mexico and Germany spent more time on applications. England's teaching materials were concerned less with reasoning than with graphic representation. On the other hand, student survey shows that they would equate reasoning with high-quality in mathematics teaching. These results from cross-national large-scale studies highlight the importance of daily classroom assessing, with which we deal in detail in the next section.

7.2 Assessing in the classroom

Two terms, Assessment *of* Learning (AoL) and Assessment *for* Learning (AfL), have been shaping discussions on how to evaluate and judge learning outcomes and learning itself since Bloom, Hastings and Madaus (1971) published their handbook on formative and summative evaluation of student learning. Sometimes from a practical perspective, AoL ends up being a shorthand for 'summative assessment', while AfL points to 'formative assessment'. But Newton (2007) argued strongly against viewing AoL as simply about results and AfL as about *use* of the results. In fact, these two are not dichotomous terms, and sometimes assessment can serve both purposes. Stiggins (2002, p. 759) argued that AfL should be conceptualized and investigated from students' perspectives, which means the purpose of AfL shifts from being pedagogy-focused, oriented towards how teachers use AfL to inform the next steps of instruction and motivate the next stages in learning, towards being learner-focused, geared towards how to encourage students to 'want to learn and feel able to learn'.

In line with this shift, at the beginning of this section, we throw light on scoring rubrics, as this type of assessment has been and continues to be overlooked. Scoring rubrics bring the assessment and learning processes together to feed off each other, and our intention is for the CCF to guide the kind of planning that takes full advantage of this.

7.2.1 Using scoring rubrics

Until now, approaches to assessing reasoning have been static, being there to assess what students have mastered and what they have not yet. But so-called 'dynamic assessment', to assess better where a student is in his or her landscape of learning, i.e. how the student progressing looks like, has been proposed as an alternative (Storeygard, Hamm and Fosnot 2010). It is also helpful at this point to consider students as assessors too, the practices in which students assess themselves and their peers. So, in this section, we will look first at scoring rubrics, followed by the use of scoring rubrics in mathematics and mathematical reasoning.

Scoring rubrics A scoring rubric consists of a set of assessment criteria, of a kind largely seen in english writing assessment. In general, there are two pairs of rubric widely used in education: analytic (versus holistic), and generic (versus task specific) (Moskal 2000). If an overall evaluation can be separated into independent factors, each of which has its own set of criteria, then an analytic rubric applies. A holistic judgement, on the other hand, is used to control and consider any overlaps in these sets of criteria. An example of generic rubric is the marking criteria for an essay or dissertation, while a task-specific one applies to a single assessment. A way to understand the impact of rubrics is to consider how they interact with student performance, as a mechanism for potential improvement. Having examined studies focused on formative uses of rubrics, Panadero and Jonsson (2013) summarized that publishing rubrics increases transparency in expectations for students and facilitates communication in the classroom.

Using scoring rubrics in mathematics Building on the holistic and analytic approaches, Turner et al. (2021) tested the idea of hybrid approaches, those combining both to assess, in this case, mathematical modelling competences at primary level. These competences were divided into four subcomponents: (1) making sense of real situations, (2) constructing mathematical models, (3) operating these models and (4) interpreting results in real situations. But the question is how to assess these subcomponents, and would the whole competence is greater than the sum of its four subcomponents? This was addressed by hybrid approach, the analytical dimension and holistic approach. Multiple-choice items were used to fulfil the analytical dimensions, while constructed response items were used for the holistic approach.

The following question is how to design the procedure of assessing. Research into designing the testing system itself consisted of three stages: (1) preliminary research, including initial capability analysis to verify the validity of content, (2) a prototype stage, which emphasized meaning in practicality, and (3) an assessment phase for both practicality and effectiveness (Plomp 2013). The component analysis is the key. For example, to assess classroom communication, Banes et al. (2020) used five factors in their observation rubric: conceptual explanations, connections between ideas, opportunities to speak, equitable participation and variety of approaches. What, then, has been used to measure *mathematical* reasoning in particular?

Using scoring rubrics in mathematical reasoning Assessing mathematical reasoning in this way is part of what makes classroom teaching and learning explicit, either in relation to its structural aspects (inductive, deductive and abductive) or its processual aspects (generalization by explanation, proof by conjecturing and justification by mathematizing), as we saw in chapter 1. This means that having a scoring rubric consisting of descriptive criteria is a way to drive the development and direction of reasoning in the classroom.

Let's look at the Australian mathematics curriculum, where reasoning is one of the four proficiency strands (with understanding, fluency and problem-solving) (Australian Curriculum, n.d., b). This curriculum describes reasoning with reference to actions such as analysing, proving, evaluating, explaining, inferring, generalizing and justifying (Australian Curriculum, n.d., a). At primary level, three types of reasoning actions are promoted: analysing, generalizing and justifying. Scholars in Australia have advocated the use of rubrics in the development of assessment methods. Loong et al. (2018) have established five levels in their generic rubric: not evident; beginning; developing; consolidating; and extending for each of the three types of reasoning sought (see Figure 7.1). The rubric encourages primary teachers to develop their understanding of the reasoning actions articulated in each category: their ability to notice the reasoning actions in the classroom, their understanding of progression in reasoning levels, and of how different actions can be connected (Bragg and Herbert 2018).

Later, Herbert et al. (2022) used this rubric in an intervention trial to examine formative assessment from teachers' perspectives. They found that this generic rubric actually turned these teachers' attention towards using the specific reasoning language from the rubric to interact with students.

Understanding scoring rubrics, as we have seen, allows insights into both the process and the structure of reasoning. The intention is to shift the teaching focus from external criteria, such as what a high-scoring performance looks like, to the best techniques to arrive at that desired outcome. Using scoring rubrics not only recognizes students' learning journey as a whole, but the rubric and learning also feed off each other. Therefore, the planning of students' learning

Figure 7.1 Assessing mathematical reasoning rubric

Student Name: **Reasoning Task:**   **Date:**

Observation of student's reasoning:		
	Analysing	**Generalising**	**Justifying**
Not Evident	• **Does not notice** common property or pattern.	• **Does not communicate** a common property or rule (conjecture).	• **Does not justify.**
Beginning	• **Recalls** random known facts or **attempts to sort** examples or **repeats** patterns.	• **Attempts to communicate** a common property or rule (conjecture) for the pattern.	• **Describes** what they did and **recognises** what is correct or incorrect. • **Argument** is not coherent or does not include all steps.
Developing	• **Notices** a common property, or **sorts and orders** cases, or **repeats and extends** patterns. • **Describes** the property or pattern.	• **Generalises:** **communicates** a rule (conjecture) using mathematical terms and records other cases or examples.	• **Attempts to verify** by testing cases and **detects and corrects** errors or inconsistencies. • Starting statements in a **logical argument** are correct.
Consolidating	• **Systematically searches** for examples, extends pattern or analyses structure to form a conjecture. • Makes **predictions** about other cases.	• **Generalises:** **communicates** a rule (conjecture) using mathematical symbols and **explains what** the rule means or **explains how** the rule works using examples.	• **Verifies** truth of statements by confirming all cases or **refutes** a claim by using a counter example. • Uses a correct **logical argument**.
Extending	• **Notices and explores** relationships between properties.	• **Generalises** cases, patterns or properties using mathematical symbols and **applies** the rule. • **Compares** different expressions for the same pattern or property to show equivalence.	• Uses a **watertight logical argument**. • **Verifies** that the generalisation holds for *all* cases using logical argument.
Comments (feedback, reasoning prompts for further development):		

Source: (Loong et al. 2018, p. 509)

journeys, including through use of the CCF, involves students in their own assessment. In the next section, we will look to the other side of the picture, exploring the challenges of using rubrics.

7.2.2 Challenges of using rubrics for formative purposes

Reaction to the idea of using rubrics itself has been diverse; inevitably, some empirical studies have been approving and others have not. Panadero and Jonsson (2020) conducted a systematic review to look at the main criticisms of using rubrics and related evidence. They found that transparency around assessment of reasoning in the classroom not only has potential to turn reasoning into a concerningly instrumental part of the curriculum (when students are told what it is and how they should do it), but also that it can possibly jeopardize creativity and divergent thinking.

Although Herbert (2019) warns that rubrics to assess reasoning are not easy to implement in the daily classroom, this has to do with lesson planning, in which reasoning elements are purposefully bedded (Davidson, Herbert and Bragg 2019). In addition, they also found that post-lesson discussions among teachers about the subtle differences between examples used in the lesson, and how they relate to the reasoning actions and levels featured in the reasoning rubric, can also feed into planning for future lessons.

Every coin has two sides. On one side, teachers' knowledge of reasoning means an ability to *notice* students' reasoning. On the other, students' ability to articulate their reasoning shows the importance of language, an easily overlooked window to the strategies that students use. Between the two, there are additional challenges, such as how to assess, track and report student progress, and the need for supporting resources, from general curriculum documents to samples of specific work (Herbert 2021). Emerging from this look at the challenges of assessing reasoning for formative purposes, the core principle is understanding how to flourish and monitor so-called high-order thinking, such as patterns of deductive reasoning (Leighton 2006).

Narrowing things down to the mathematics subject at secondary level in England, the concepts covered are more formal and abstract, so AfL tends to take the form of an algorithm, using a procedural approach and a strong element of justification (Hodgen and Marshall 2005). This study also argued that mathematics teachers should pay more attention to 'the role of language within thinking and the role of dialogue in learning', just as teachers of english in secondary schools do (Hodgen and Marshall 2005, p. 173). Bearing this in mind, we now look at experiences of building up reasoning assessment.

7.3 Does the reasoning experience count?

This section takes an entirely different approach from AfL and AoL, both of which are concerned with the end game, what to assess. Instead, it offers ways to put the focus on the *process* or the *experience* of reasoning. We start from two points raised in research: assessment within teaching practice and students' learning experience, then moving on to look at how to assess (creative) reasoning in reality.

7.3.1 Teacher assessment

There is an impressive array of research projects sharing a concern for reasoning, this narrow subtopic of the testing system. More generally, testing, especially in national assessments, serves as a distillation of assumptions embedded in society, concerning the quality of teaching and learning and school accountability. The Cockcroft report (Cockcroft 1982) advocated for teacher assessment to be included among the methods of assessment in public examinations, especially as they relate to communication, reasoning and problem-solving. As such, assessing reasoning can break free of the impoverished view of testing that equates it to pencil-and-paper tasks. Instead, it can be integrated as ongoing assessment, in a way that may overcome concerns expressed by Ollerton and Vasile (2014) about the dangers of separating assessment from classroom practice.

Lewis (2015, p. 22) argued that assessing reasoning is 'a significant challenge' in the daily classroom, because it potentially includes two dimensions: assessing reasoning per se as a cognitive function, and assessing students' reasoning experiences. Both require careful consideration of how reasoning can be fostered through the sequencing of relevant tasks. This can lead to the dilemma of whether to assess reasoning for the sake of reasoning or for the sake of checking knowledge and skills.

Palm, Boesen and Lithner (2011) investigated the differences between types of reasoning required in national tests compared with tests made by teachers, in the context of the Swedish upper-secondary level. The analytical framework they proposed divides test tasks into those prompting imitative reasoning and those needing reasoning that is creative and mathematically founded (which we discussed in chapter 1 and earlier in this chapter). Although the Swedish national curriculum features different levels of alignment, and despite participating teachers having different levels of experience in making tests, there emerged a clear general finding that teacher-made tests were more geared towards imitative reasoning. This also points to a need for teachers to be support and given access to professional development in this area, something which we will discuss in the next chapter.

However, in the context of a mathematics classroom, reasoning is not entirely separate from understanding – either relational or procedural understanding. Compared with that in classrooms in Flanders, Hungary, and Spain, English practice tends to focus on low-level procedural expectations, which subordinates higher-level learning (Andrews and Sayers 2008). This higher-level learning, as argued by Foster (2021), may be enabled by using confidence assessment – asking students how sure they are about their answer – which leads to self-checking, self-explanation and reasoning itself.

7.3.2 Tasks to build up reasoning

The reality of the daily classroom is that reasoning starts from the solving of tasks, both routine and non-routine ones – and the same is true in exams. Lithner (2000) uses the term 'reasoning' – the product of students' thinking to reach conclusions – along with 'argumentation' to summarize the process of thinking. Based on this understanding of the thinking process, a four-step reasoning structure is proposed: (1) problematic situation, (2) strategy choice, (3) strategy implementation, and (4) conclusion. Within this structure, in the task-solving process, argumentation seem to equate to generalization by explanation and to justification by mathematizing, as discussed in chapter 1, rather than to logical proof.

Lithner (2000, 2004) went on to propose the ideas of 'plausible reasoning' and 'experienced reasoning', which are also both part of this four-step structure. Plausible reasoning is reasoning based on mathematics properties, while experienced reasoning is that based on procedures and notions from students' previous learning experiences. This research reveals that experienced reasoning is the dominant strategy choice on the global level, on the level of selecting and implementing strategies. Plausible reasoning follows on at a local level, but in a limited way. Yet these insights do not provide a definitive answer to the questions still surrounding how, in reality, reasoning can best be assessed. We argue that, at the confluence of plausible reasoning and experienced reasoning, high-level learning and low-level procedural expectation, there lies an important contribution of the CCF. Our framework aims at maximizing how reasoning is built in from the lesson planning stage, through the creation of space for reflection.

Let's consider the example of multiplicative reasoning, which is the main cognitive shift students face in lower-secondary settings. Multiplicative reasoning is linked to concepts such as ratio, numbers (decimals, fractions) and algebra, but it has been noted that teaching outcomes in this area are not particularly positive (Brown, Küchemann and Hodgen 2010). The Increasing Competence and Confidence in Algebra and Multiplicative Structures (ICCAMS) project provided formative assessment guidance to empower the implementation of multiplicative reasoning. Materials created on multiplicative reasoning include a three-block lesson structure, which looks at models of multiplication, fractions and identifying multiplicative relations, respectively (ICCAMS, n.d.). The project provides a package of teaching materials which includes explanations of mathematical ideas, information about students' potential mathematical experiences, key questions for the lessons, and descriptions of which tasks assess which aspect of mathematical reasoning and in which way (such as peer- and self-assessment). It seems that reasoning-oriented tasks can be the basis for assessing reasoning, and something upon which all lesson planning can be built.

What we face is not a matter of assessing reasoning, but of lesson planning that integrates reasoning into our recognition of the nature of mathematics, and into how tasks are structured in a logical way. It is not necessary to distinguish learning from performing, and to separate teaching from assessing. It is better, simply, to begin lesson planning from the principles of the CCF so that mathematical reasoning is properly embraced at all levels. The next question arising, then, is about 'tasks': what sort of tasks count as reasoning-related, according to this understanding.

7.4 What comes next?

In this chapter, we have explored the two different ways of assessing reasoning, as emerging from different views of reasoning itself: structured intervention studies that work from a view of reasoning as a product, and semi-structured comparative judgement studies that view reasoning as a process. These two ways are both surrounded by various levels of contextual factors, to do with what different education systems offer and expect regarding reasoning. This is the basis for our argument that linking assessing with classroom practice is not enough. Assessing should be embedded into the lesson planning stage and within every task if it is to be properly built into students' reasoning experiences. Lesson planning matters for students' experiences, and our framework for lesson planning connects closely to these reasoning experiences. So, in the next chapter, we turn to ways in which school leaders can integrate this into practice.

Practice activities

Chapter 6 asked for the criteria used to judge which narrative about trigonometry is the best. Now reflect on your assessing process and, ideally, compare your criteria with those used by other people. Then summarize the underlying principles of your own assessment criteria, i.e. your beliefs about conceptual understanding in trigonometry.

References

Ainley, J. and Carstens, R. (2019) Teaching and Learning International Survey (TALIS) 2018 conceptual framework. OECD Education Working Papers No. 187. Available at: https://www.oecd-ilibrary.org/docserver/799337c2-en.pdf?expires=1649662724&id=id&accname=guest&checksum=14FEB14F9F5C9F2AD2A9A51F44E29182 (accessed 12 June 2023).

Andrews, P. and Sayers, J. (2008) Conditions for learning: Part 4. *Mathematics Teaching*, 209: 10–13. Available at: https://eric.ed.gov/?id=EJ815279 (accessed 6 September 2023).

Australian Curriculum (n.d., a) *Reasoning*. Available at: https://www.australiancurriculum.edu.au/resources/mathematics-proficiencies/portfolios/reasoning/ (accessed 12 June 2023).

Australian Curriculum (n.d., b) *Structure*. Available at: https://www.australiancurriculum.edu.au/f-10-curriculum/mathematics/structure/ (accessed 12 June 2023).

Banes, L.C., Restani, R.M., Ambrose, R.C. et al. (2020) Relating performance on written assessments to features of mathematics discussion, *International Journal of Science and Mathematics Education*, 18(7): 1375–98. https://doi.org/10.1007/s10763-019-10029-w

Basturk, S. (2010) First-year secondary school mathematics students' conceptions of mathematical proofs and proving, *Educational Studies*, 36(3): 283–98. https://doi.org/10.1080/03055690903424964

Bisson, M.J., Gilmore, C., Inglis, M. and Jones, I. (2016) Measuring conceptual understanding using comparative judgement, *International Journal of Research in Undergraduate Mathematics Education*, 2(2): 141–64. https://doi.org/10.1007/s40753-016-0024-3

Bloom, B.S., Hastings, J.T. and Madaus, G.F. (1971) *Handbook on formative and summative evaluation of student learning.* New York: McGraw-Hill.

Bragg, L.A. and Herbert, S. (2018) What can be learned from teachers assessing mathematical reasoning: a case study. *Mathematics Education Research Group of Australasia.* Available at: https://files.eric.ed.gov/fulltext/ED592480.pdf (accessed 28 June 2023).

Brodie, K. (2009) *Teaching mathematical reasoning in secondary school classrooms* Vol. 775. Berlin/Heidelberg, Germany: Springer Science & Business Media.

Brown, M., Küchemann, D. and Hodgen, J. (2010) *The struggle to achieve multiplicative reasoning 11-14*, in M. Joubert. and Andrews, P. (eds.) British Congress for Mathematics Education (BCME-7), University of Manchester, Manchester, England. Available at: https://iccams-maths.org/wp-content/uploads/2016/01/BCME-2014-The-struggle-to-achieve-multiplicative-reasoning-11-14.pdf (accessed 3 July 2023).

Cockcroft, W. (1982) Mathematics counts, report of the committee of Inquiry into the Teaching of mathematics in schools under the Chairmanship of Dr WH Cockcroft. London: Her Majesty's Stationery Office. Available at: https://www.educationengland.org.uk/documents/cockcroft/cockcroft1982.html (accessed 12 June 2023).

Davidson, A., Herbert, S. and Bragg, L.A. (2019) Supporting elementary teachers' planning and assessing of mathematical reasoning, *International Journal of Science and Mathematics Education*, 17(6): 1151–71.

Davies, B., Alcock, L. and Jones, I. (2020) Comparative judgement, proof summaries and proof comprehension, *Educational Studies in Mathematics*, 105(2): 181–97. https://doi.org/10.1007/s10649-020-09984-x

Davies, B., Alcock, L. and Jones, I. (2021) What do mathematicians mean by proof? A comparative-judgement study of students' and mathematicians' views, *The Journal of Mathematical Behavior*, 61: 1–10. https://doi.org/10.1016/j.jmathb.2020.100824

Foster, C. (2021) Implementing confidence assessment in low-stakes, formative mathematics assessments, *International Journal of Science and Mathematics Education*. 20: 1411–29. https://doi.org/10.1007/s10763-021-10207-9

Grossman, J. and Mackenzie, F. J. (2005) The randomized controlled trial: gold standard, or merely standard?, *Perspectives in biology and medicine*, 48(4): 516–34. Available at: https://muse.jhu.edu/article/188201 (accessed 6 September 2023).

Herbert, S. (2019) Challenges in assessing mathematical reasoning. *Mathematics Education Research Group of Australasia.* Available at: https://files.eric.ed.gov/fulltext/ED604186.pdf (accessed 12 June 2023).

Herbert, S. (2021) Overcoming challenges in assessing mathematical reasoning. *Australian Journal of Teacher Education (Online)*, 46(8): 17–30. Available at: https://ro.ecu.edu.au/cgi/viewcontent.cgi?article=4665&context=ajte (accessed 12 June 2023).

Herbert, S., Vale, C., White, P. and Bragg, L.A. (2022) Engagement with a formative assessment rubric: a case of mathematical reasoning, *International Journal of Educational Research*, 111: 1–17. https://doi.org/10.1016/j.ijer.2021.101899

Hodgen, J. and Marshall, B. (2005) Assessment for learning in English and mathematics: a comparison, Curriculum journal, 16(2): 153–76. https://doi.org/10.1080/09585170500135954

ICCAMS (n.d.) Multiplicative reasoning. Available at: http://iccams-maths.org/multiplicative-reasoning/ (accessed 12 June 2023).

Jones, I. and Inglis, M. (2015) The problem of assessing problem solving: can comparative judgement help?, *Educational Studies in Mathematics*, 89: 337–55. https://doi.org/10.1007/s10649-015-9607-1

Leighton, J.P. (2006) Teaching and assessing deductive reasoning skills, *The Journal of experimental education*, 74(2): 107–36. https://doi.org/10.3200/JEXE.74.2.107-136

Lewis, M. (2015) Assessment: beyond right and wrong, *Mathematics Teaching*, 249: 21–23. Available at: https://www.atm.org.uk/Mathematics-Teaching-Journal-Archive/83727 (accessed 6 September 2023).

Lithner, J. (2000) Mathematical reasoning in school tasks, *Educational Studies in Mathematics*, 41(2): 165–90. Available at: https://link.springer.com/content/pdf/10.1023/A:1003956417456.pdf (accessed 12 June 2023).

Lithner, J. (2004) Mathematical reasoning in calculus textbook exercises, *The Journal of Mathematical Behavior*, 23(4): 405–27. https://doi.org/10.1016/j.jmathb.2004.09.003

Loong, E., Vale, C., Widjaja, W. et al. (2018) Developing a Rubric for Assessing Mathematical Reasoning: A Design-Based Research Study in Primary Classrooms, In Hunter, J., Perger, P. and Darragh, L. (eds.). *Making waves, opening spaces (Proceedings of the 41st annual conference of the Mathematics Education Research Group of Australasia)*: 503–10. Auckland: MERGA. Available at: https://files.eric.ed.gov/fulltext/ED592421.pdf (accessed 12 June 2023).

Lowrie, T., Harris, D., Logan, T. and Hegarty, M. (2019) The impact of a spatial intervention program on students' spatial reasoning and mathematics performance, *The Journal of Experimental Education*, 89(2): 259–77. https://doi.org/10.1080/00220973.2019.1684869

Lowrie, T., Logan, T., Harris, D. and Hegarty, M. (2018) The impact of an intervention program on students' spatial reasoning: Student engagement through mathematics-enhanced learning activities, *Cognitive research: principles and implications*, 3(1): 1–10. https://doi.org/10.1186/s41235-018-0147-y

Moskal, B.M. (2000) Scoring rubrics: what, when and how?, *Practical Assessment, Research, and Evaluation*, 7(1): 3. https://doi.org/10.7275/a5vq-7q66

Newton, P.E. (2007) Clarifying the purposes of educational assessment, *Assessment in education*, 14(2): 149–70. https://doi.org/10.1080/09695940701478321

Nunes, T. and Csapó, B. (2011) Developing and assessing mathematical reasoning. Available at: http://publicatio.bibl.u-szeged.hu/11244/1/Math_Framework_English_17_56_u.pdf (accessed 11 June 2023).

OECD (2009) *Creating effective teaching and learning environments*. Available at: https://www.oecd.org/education/school/43023606.pdf (accessed 12 June 2023).

OECD (2014) TALIS 2013 Results: An International Perspective on Teaching and Learning. TALIS. Place missing: OECD Publishing. http://dx.doi.org/10.1787/9789264196261-en

OECD (2016) *Ten Questions for Mathematics Teachers ... and how PISA can help answer them*. PISA. Paris: OECD Publishing. https://doi.org/10.1787/9789264265387-en

OECD (2018) PISA 2021 Mathematics Framework (Draft). Available at: https://www.oecd.org/pisa/pisaproducts/pisa-2021-mathematics-framework-draft.pdf (accessed 12 July 2023).

OECD (2020) Global Teaching InSights: A video study of teaching. Paris: OECD Publishing. https://doi.org/10.1787/20d6f36b-en

Ollerton, M. and Vasile, D. (2014) 'Fluffy' assessment, joined-up thinking, vision and future aspiration, *Mathematics Teaching*, 238: 11–13. Available at: https://www.atm.org.uk/MT-Tasters-Materials- (accessed 6 September 2023).

Palm, T., Boesen, J. and Lithner, J. (2011) Mathematical reasoning requirements in Swedish upper secondary level assessments, *Mathematical Thinking and Learning*, 13(3): 221–46. https://doi.org/10.1080/10986065.2011.564994

Panadero, E. and Jonsson, A. (2013) The use of scoring rubrics for formative assessment purposes revisited: A review, *Educational research review*, 9: 129–44. https://doi.org/10.1016/j.edurev.2013.01.002

Panadero, E. and Jonsson, A. (2020) A critical review of the arguments against the use of rubrics, *Educational Research Review*, 30: 1–19. https://doi.org/10.1016/j.edurev.2020.100329

Plomp, T. (2013) Educational Design Research: An Introduction, in T. Plomp and N. Nieveen (eds), *Educational Design Research Part A: An Introduction*. Enschede, The Netherlands: SLO.

Pollitt, A. (2012) The method of adaptive comparative judgement, *Assessment in Education: Principles Policy and Practice*, 19(3): 281–300. https://doi.org/10.1080/0969594X.2012.665354

Stiggins, R.J. (2002) Assessment crisis: the absence of assessment for learning, *Phi Delta Kappan*, 83(10): 758–65. https://doi.org/10.1177/003172170208301010

Storeygard, J., Hamm, J. and Fosnot, C.T. (2010) Determining what children know: Dynamic versus static assessment, in National Council of Teachers of Mathematics (eds) *Models of intervention in mathematics: Reweaving the tapestry*. Reston, VA: NCTM.

Suurtamm, C., Thompson, D.R., Kim, R.Y. et al. (2016) *Assessment in mathematics education: Large-scale assessment and classroom assessment*. London: Springer Nature. http://doi.org/10.1007/978-3-319-32394-7_1

Turner, E.E., Chen, M.K., Roth McDuffie, A. et al. (2021) Validating a student assessment of mathematical modeling at elementary school level, *School Science and Mathematics*, 121(7): 408–21. https://doi.org/10.1111/ssm.12494

8 So what next? The critical role of school leaders in making this happen

Key arguments

In this final chapter, we argue that, while it is school leaders who typically oversee changes to curricula and pedagogy in their schools, they don't necessarily have the time or the skills to introduce new initiative themselves. As such, to help you take on board some of the maths-related ideas, approaches and frameworks we present, it is useful to consider the idea of who might support their introduction. We refer to the individuals who can help in this way as 'change agents'. Change agents can be identified in a number of ways. For instance, their position within a school's social network. Likewise, change agents may be those best able to 'signal' that a specific change is attractive enough for others to adopt through championing the change. Effective change agents also often benefit from 'homophily' – they are those that others clearly gravitate towards. Thinking about how these aspects translate in the staffroom, means we might select change agents based on their subject mastery, attitudes to the change in question, their standing with colleagues and so on. They must also, of course, have the ability to lead change: this requires school leaders to ensure change agents are supported effectively.

Key terms

- Change agent
  Those individuals situated within a school, best able to catalyse the successful introduction of innovations or perspectives to teaching colleagues and others.
- Social networks
  A social network represents a set of relevant actors (persons or groups) connected to each other by a specific type of relationship, which enable individuals to access a range of practitioner-based social capital resources
- Homophily
  The idea that like attracts like or, as it's more commonly expressed, that 'birds of a feather flock together'. In other words, our tendency to interact with others who are like us.

The process of getting teachers engaged with new ideas and encouraging them to trial these innovations in the classroom is complicated. The CCF may initially seem a little remote from everyday school management processes, from deciding on school directions, from promoting staff motivation and commitment and from the success of a schools. But it does not have to be this way. In the first section, we discuss the first- and second-order effects of deploying the CCF for school leaders. Based on this, we propose how to support the change.

8.1 Rethinking the role of school leaders

School leaders have a substantive role in improving outcomes for children and young people (e.g. Marzano, Waters and McNulty 2005; Robinson, Hohepa and Lloyd 2008; Robinson, Lloyd and Rowe 2008). In fact, in terms of *within-school* factors, their impact is second only to that of teachers (Leithwood and Louis 2012). School leaders are able to make a difference to teaching and learning though what are known as first- and second-order effects. Targeting first-order variables may mean, for instance, using instructional leadership to improve the quality of teaching and the nature of the curriculum delivered to students in the classroom (Tulowitzki and Pietsch 2018). An example of generating second-order effects is using transformational leadership to increase the commitment of others in the school in relation to specific first-order effects on learning (Tulowitzki and Pietsch 2018).

But, despite their latent ability to do so, school leaders do not necessarily have the time or skills to introduce new initiatives, such as the CCF, themselves. Hence a vital leadership skill is knowing how to delegate: empowering others to deliver for them. When it comes to education, therefore, understanding who might be able to lead change in your school is vital. As such, its useful to think of people who might support the introduction of new initiatives as 'change agents' – those in a school with the means (typically capacity and opportunity) to successfully transform aspects of how the school operates (Fullan 2011).

Change agents, in other words, are the educators best able to catalyse the successful introduction of innovations or perspectives to teaching colleagues and others. Change agents are increasingly viewed as vital to the successful operation of schools and school systems. For instance, in 'self-improving' school systems, such as those in England, Ontario and New South Wales, improvements in children's outcomes are positioned as occurring when teachers mobilize innovations, practices, perspectives and ideas (collectively described as 'new ways of working') amongst colleagues (Ainscow 2014; Greany and Higham 2018). As these new ways of working are adopted, the attitudes and practices of teachers and other practitioners change, ideally resulting in improvements in student outcomes (academic or otherwise: Earley and Greany 2017). The key agents for change in the case of the CCF is, then, not necessary the head of maths.

In self-improving school systems, the mobilization of new ways of working is undertaken by teachers and informal leaders as well as formal school leaders

(Kotter 2014; Wenner and Campbell 2017). Yet not all teachers are equal in their ability to mobilize new ways of working, such that they are adopted widely. Understanding which educators are best able to encourage the take up of new ways of working is therefore vital to ensuring that the change being sought actually makes a difference – in other words, that it is able to continuously improve the education students receive, as well as improving equity in student achievement.

So, who makes an effective educational change agent? Findings from a recent systematic review by Brown, White and Kelly (2021) provide some useful insight here. For example, when viewed through the lens of *social network theory*, the success or failure of educational change is dependent on the social networks through which it is mediated (Coburn et al. 2010; Warren Little 1990). A social network represents a set of relevant actors (persons or groups) connected to each other by a specific type of relationship that enables individuals to access a range of practitioner-based social capital resources (Daly 2010). For instance, 'instrumental' social capital resources such as information sharing, advice giving and problem-solving provide concrete support for achieving specific goals. In contrast, 'expressive' social capital refers to resources such as trust, support and encouragement, all of which can influence attitudes towards given goals and instil the resilience required to keep pursuing them (Puccia et al. 2021). From this perspective, change agents will be those individuals best situated within a social network to mobilize both types of social capital in support of a given change (Battilana and Casciaro 2013).

From another perspective (that of 'organizational semiotics': e.g. Gazendam, Jorna and Cijsouw 2003), effective change agents are those who are best able to 'signal' that a specific change is attractive enough for others to adopt. In this sense, 'attractiveness' can refer to the idea represented by the change in question, but it can also represent the extent to which an idea appears *achievable*; in other words, whether those expected to change believe they possess the ability to successfully do so. As with any form of semiotic, a 'thing' (an object or idea) only has meaning when viewed in relation to other 'things' (Eco 1979). A change agent can therefore position a change as attractive by contrasting it with something that teachers regard as less attractive. For example, Schildkamp and Datnow (2020) observe that teachers are far *more* likely to consider using data to inform their practice when there is an explicit focus on equity, than when data use is undertaken in the service of accountability.

The achievability of the change therefore involves change agents signalling certain *attributes* of the change in question; for instance, how easy a change is to master, and/or the extent to which it involves changers drawing on familiar sets of skills and practices. According to Rogers (1995), these attributes concern five key aspects of an innovation: 1) its relative advantage; 2) its compatibility; 3) its complexity; 4) its observability; and 5) its trialability. Detail on each of these is set out in Table 8.1, with adoption much more likely to occur when these characteristics are addressed.

Table 8.1 Five attributes affecting the adoption of innovations

Attribute	Conceptual definition
Relative advantage	The extent to which an innovation is perceived as better than the idea it supersedes
Compatibility	The degree to which an innovation is perceived as being consistent with the existing values, past experiences and needs of potential adopters
Complexity	The extent to which an innovation is perceived as difficult to use
Observability	How visible the results of an innovation are to others
Trialability	Whether and the extent to which an innovation may be experimented with on a limited basis

Source: (taken from Rogers 1995, pp. 15–16)

There are also psychological perspectives that add insight; specifically, the concept of heuristics (Kahneman 2011; Tversky and Kahneman 1974). One common heuristic is homophily: the idea that like attracts like or, as it is more commonly expressed, that 'birds of a feather flock together'. The notion of homophily can be found as far back as in Ancient Greece, where Socrates is reported to have suggested that 'similarity begets friendship' and as such, we should 'delight in equals'. Aristotle likewise observed that we tend to 'love those who are like ourselves'. But as a phenomenon, our tendency to interact with others who are like us was only labelled 'homophily' by Paul Lazarsfeld and Robert Merton in 1954, with the etymology of the word deriving from the combination of two Ancient Greek words: *homo* meaning 'the same' and *–phily* meaning 'liking' (Lazarsfeld and Merton 1954).

There is some debate as to whether homophily is an objective preference, but that notwithstanding, what the heuristic perspective suggests is that change agents are those most able to galvanize change because they are viewed as people who are acceptable to follow: they are seen as being 'like me'; as possessing admirable qualities; they are likely to be charismatic; are connected to 'others whom I like'. The most effective change agents are thus those that others clearly gravitate towards.

These three lenses – the social network, the semiotic and the heuristic – are most useful when considering change agents as those attempting to influence an organization *from the bottom up*. In other words, they can help identify who best to set free on the ground, with the purpose of galvanizing change in relation to new maths programmes; seeding them with the ideas within this book and letting them get on with it. Further attributes of effective 'bottom-up' change agents are that:

1 They display **agency**: change agents evaluate need and activate change through a collaborative process that attends to the motivation of others (Lai and Cheung 2014; Lukacs 2012; Wenner and Campbell 2017);

2 They display **cultural competence**: specifically, they are aware of the sociocultural context they operate in, have high expectations, a desire to make a difference, and are cognizant of the need to challenge the deficit mindset of colleagues. This type of change leader may also identify means through which to overcome the professional antinomies (contradictions) often faced by teachers working in disadvantaged and challenging situations. They may draw on those holding 'local knowledge', such as teaching assistants (Hauge, Norenes and Vedøy 2014; Lee and Louis 2019; Von Hippel 2014);

3 They are **effective relationship builders** with colleagues within their school, as well as those external to it. Change agents engage with key local stakeholders (parents, community groups and so forth) to co-construct the difference they are seeking to achieve and the means of achieving it (Schnellert 2020; Poekert, Alexandrou and Shannon 2016).

Now, who would be the potential candidates in mind for CCF change agents? Make sure they develop their own understanding of CCF and a personalized way to adapt the lesson planning approach, before working with other practitioners who might be resistant or confused in some way.

8.2 Finding others to work with change agents

It is useful to seek out other allies to back up the endeavours of change agents. To work out who might be most effective in support, we can examine what we currently know about which teachers are best placed to lead *top-down* change. This applies especially when the nature of the change in question has more or less already been determined (either by school leaders or by the bottom-up change agents just discussed). An in-depth review of this area undertaken by Brown, White and Kelly (2021) suggests that choosing 'top down' change agents based on their possession of specific qualities can be effective. These qualities concern areas including:

1 **Attitudes to change** generally or to a particular change, including knowledge, beliefs and values (Fullan 2011; Poekert, Alexandrou and Shannon 2016).

2 **Mastery of subject knowledge** and/or pedagogy; for instance, whether they possess expertise, which in itself was viewed as a function of years of experience and perceived subject knowledge (Booth et al. 2021).

3 Whether the change agent is a **lifelong learner**: someone who is curious, open-minded or has a growth mindset and is willing to try new approaches (Ali 2011; Beauchamp 2015; Watson 2014).

4 Whether the change agent has **entrepreneurial qualities**; for instance, whether they are happy to take risks to see if a change can be enhanced further. Similarly, whether they can encourage others to do the same (OECD 2016; Wenner and Campbell 2017).

5 Whether the change agent is an **effective collaborator**, with strong colle-
gial standing, and someone who can leverage their networks effectively
to help secure change (Battilana and Casciaro 2013; Hairon and Goh 2015;
Warren Little 1990).

We can also consider which educators might comprise effective change agents
by thinking about the roles they will undertake. Here research by Lai and
Cheung (2014) suggests that the following six roles are involved in introducing
top-down change:

1 with other school members around school reform efforts
2 striving for pedagogical excellence
3 confronting barriers in the school's culture and structures
4 translating ideas into actions
5 participating in decision-making
6 taking the initiative in leading school improvement.

Concerning (1) interaction, change agents need to know – in depth – the types
of people they are likely to be dealing with. Of course, not everyone is likely
to embrace innovation. Rogers (1995) classifies individuals according to their
relationship with innovation-related change: 1) innovators; 2) early adopters;
3) early majority; 4) late majority; or 5) laggards. The definitions of these
adopter types are set out in Table 8.2:

Table 8.2 Adopter types identified by Rogers

Adopter type	Definition	% of population
(1) Innovators	When it comes to new ideas, innovators are active information-seekers. They have a high degree of mass media exposure and their interpersonal networks extend over a wide area, reaching outside of their organization. Furthermore, the innovator plays an important role in the launching of new ideas by importing them from outside their organization's boundaries.	2.5
(2) Early adopters	Early adopters are a more integrated part of the local system than innovators and have the greatest degree of opinion leadership. The early adopter is considered by many 'the individual to check with' before using a new idea and serves as a role model for the rest of the organization.	13.5

(continued)

Table 8.2 *(Continued)*

Adopter type	Definition	% of population
(3) Early majority	Members of the early majority adopt new ideas just before the average member of the organization, but seldom hold positions of leadership. The early majority typically deliberate for some time before completely adopting a new idea.	34.0
(4) Late majority	Members of the late majority adopt new ideas just after the average member of the organization. Innovations are approached with a sceptical and cautious air, and the late majority do not adopt until most others have done so. Furthermore, almost all of the uncertainty about a new idea must be removed before members of the late majority feel it is safe to adopt.	34.0
(5) Laggards	Laggards are the last of the organization to adopt an innovation. Typically, the decisions of laggards are often made in relation to what has been done previously and laggards tend to be suspicious of innovation and change.	16.0

Source: (Rogers 1995, pp. 264–5)

According to Rogers, then, the vast majority of individuals adopt innovations only after someone else has already done so. This reinforces the perspectives discussed above: that change agents are those individuals best situated within a social network to mobilize the types of social capital needed to ensure a given change happens. More specifically, however, this means we need to select change agents according to the level of social influence they have within a network. As we all know, social influence can have a material impact on people's attitudes and behaviours; in other words, our choices and decisions and our opinions and beliefs are, more often than not, influenced by others (Berger 2016). But it is also clear that social influence can assert itself a number of ways, including:

1 Through implicit norms and guidelines that govern our understanding of how to respond in specific situations (Berger 2016).

2 Through individuals relying on the judgement of others when they are uncertain, meaning that the views within groups can converge (Asch 1956).

3 When individuals use the behaviour of others as a source of information to guide them in how to act – or as Berger observes, as 'a heuristic that simplifies decision making' (2016, p. 29) (Berger provides myriad examples to

illustrate this point, ranging from where we park our car to how we decide which school to send our children to).

4 As people often feel social pressure to conform with the decisions or behaviour of the wider group (Berger 2016).

So, having on board a change agent who possesses social influence opens up numerous approaches to influencing the take up of new innovations, from having the change agent explicitly promote the idea, to actively using it and being seen to actively use it. Or, as Lai and Cheung (2014) argue, by attempting to achieve three goals:

1 Encouraging others to improve their professional practice.
2 Nurturing a culture of success.
3 Continuously demonstrating professionalism (i.e. 'walking the talk').

So again, those selected as change agents need to be capable of achieving these goals and carrying out these tasks. At the same time, however, current research fails to articulate the specific actions and tactics that top-down change agents should adopt as they engage in those relationships and interactions so as to effectively change the pedagogy of other teachers. Research into distributed leadership, (e.g. Brown, MacGregor and Flood 2020), however, indicates that distributed leaders can work effectively as agents of change when they lead processes of professional inquiry as part of in-school professional learning communities. This points specifically to when distributed leaders attempt to:

1 Introduce change by guiding their colleagues to explore specific issues of teaching and learning.
2 Introduce colleagues to new ideas relating to specific problems of teaching and learning.
3 Support colleagues in testing out these new ideas in risk-free environments.
4 Invite colleagues to consider the impact of new approaches to teaching and learning and how they can be refined, augmented and incorporated into existing practice.

8.3 What support do change agents need?

Yet once change agents have been identified, there are a number of factors that can enable or inhibit their ability to succeed. These include: (1) principal or school leadership support; (2) buy-in to the change agent role among practitioner colleagues; (3) access to training and professional development; and (4) perceived autonomy and the change agent's own positioning in the role (Poekert, Alexandrou and Shannon 2016). To begin with, school leaders can show acknowledgment and recognition of teachers' roles as change agents by providing them with classroom release time to work with colleagues, remuneration for the role or other organizational support (Brown and Flood 2019).

A lack of time or structural resourcing has been noted as a major barrier to a change agent's work, especially since their work is likely to be additional to already busy teaching workloads (Brown 2020; Brown and Flood 2019). Regarding the second factor, teacher resistance to change can make the work of change agents difficult, and a perceived lack of support from school leaders can fuel such resistance (Brown, White and Kelly 2021). Again, school leaders can provide assistance here, since a shared vision for change can improve a change agent's positioning in the eyes of staff members (Brown, White and Kelly 2021).

There seems to be little preparation and training afforded to teachers to act as change agents. Programmes addressing the topic range widely from conferences to centralized professional development activities and from local training courses to university master's degrees (Wenner and Campbell 2017). Findings from a rapid literature review undertaken by Booth et al. (2021) include the suggestion that 'second stage' teachers (defined as those with three to ten years of experience) who take on 'reform' roles (roles which involve attempting to change the practice of colleagues) generally benefit from two forms of professional development.

The first type is that which helps them promote their role, especially when norms exist within schools regarding teacher autonomy and when respect tends to be reserved for those with the highest levels of experience and seniority. The second form of professional development is that designed to support change agents when they encounter resistance to change in their context. Change agents' own perceptions of their role, and their autonomy to act, can either support or hinder them in fulfilling their mission. The change agent's role is often perceived as blurring the line between teaching and leadership in schools and research has found that, as such, teachers can struggle to define and identify with it (Poekert, Alexandrou and Shannon 2016).

Furthermore, the process evaluation of the Research Learning Network intervention (Rose et al. 2017), where opinion-formers were used as change agents, found that staff turnover, competing priorities and teachers' time limitations were barriers to the successful implementation of change, echoing some of the issues noted above. Similarly, work undertaken by Brown and Flood (2019) indicates that if distributed leaders are to be effective change agents, then school leaders need to attend to three areas. First, school leaders need to ensure that distributed leadership activity is formally linked to the policies and processes of the school (such as by school improvement plans). Doing so signals its importance, and positions such activity as something that is key to a school's culture and way of working. Second, time and space needs to be created for distributed leaders to interact with colleagues, thus enabling new ideas to be mobilized. Third, distributed leaders need to be supported in understanding how best to mobilize new ideas. This is particularly important, given that our current understanding indicates that the passive dissemination of new ideas and practices is ineffective, while the most impactful forms of mobilization involve school staff collaboratively engaging with innovations.

So, after fully introducing the CCF outlined in this book, we leave three questions: 1) who needs to introduce change?; 2) why?; and, perhaps most importantly, 3) what to do to support them to lead change effectively or what support do you need?

Practice activities

Can you identify the change agents in your school? How might you: 1) support such change agents? 2) Ensure practitioner colleagues buy into their role; and 3) ensure they have access to training and professional development to help them in their role?

References

Ainscow, M. (2014) *Towards self-improving school systems lessons from a city challenge.* Abingdon, England: Routledge.

Ali, T. (2011) Understanding the evolving roles of improvement-oriented high school teachers in Gilgit-Baltistan, *The Qualitative Report*, 16(6): 1616–44. https://doi.org/10.46743/2160-3715/2011.1320

Asch, S. (1956) Studies of independence and conformity: A minority of one against a unanimous majority, *Psychological Monographs: General and Applied*, 70(9): 1–70. https://doi.org/10.1037/h0093718

Battilana, J. and Casciaro T. (2013) The network secrets of great change agents, *Harvard Business Review*, 91: 7–8, 62–8, 132. Available at: https://mycourses.aalto.fi/pluginfile.php/832376/course/section/129609/Battilana_Casciaro_The%20network%20secrets%20of%20great%20change%20agents_HBR.pdf (accessed 25 September 2023)

Beauchamp, C. (2015) Reflection in teacher education: issues emerging from a review of current literature, *Reflective Practice*, 16(1): 123–41. https://doi.org/10.1080/14623943.2014.982525

Berger, J. (2016) *Invisible influence: the hidden forces that shape behaviour.* New York: Simon & Schuster.

Booth, J., Coldwell, M., Müller, L.-M. et al. (2021) Mid-career teachers: A mixed methods scoping study of professional development, career progression and retention, *Education Sciences*, 11: 299. http://dx.doi.org/10.3390/educsci11060299

Brown, C. (2020) *The Networked School Leader: How to improve teaching and student outcomes using learning networks* London: Emerald.

Brown, C. and Flood, J. (2019) *Formalize, prioritize and mobilize: How school leaders secure the benefits of professional learning networks.* London: Emerald.

Brown, C., MacGregor, S. and Flood, J. (2020) Can models of distributed leadership be used to mobilize networked generated innovation in schools? A case study from England, *Teaching and Teacher Education*, 94. Available at: https://www.sciencedirect.com/science/article/abs/pii/S0742051X19313782?via%3Dihub (accessed 15 June 2023).

Brown, C., White, R. and Kelly, A. (2021) Teachers as educational change agents: what do we currently know? Findings from a systematic review [version 1; peer review:

2 approved], *Emerald Open Research 2021*, 3: 26. https://doi.org/10.35241/emeral-dopenres.14385.1

Coburn, C.E., Choi, L. and Mata, W.S. (2010) "I would go to her because her mind is math": Network formation in the context of district-based mathematics reform, in A.J. Daly (ed.) *Social Network Theory and Educational Change*. Boston: Harvard Education Press.

Daly, A. (2010) Mapping the terrain: Social network theory and educational change, in A.J. Daly (ed.) *Social Network Theory and Educational Change*. Boston: Harvard Education Press.

Earley, P. and Greany, T. (2017) The future of leadership, in P. Earley and T. Greany (eds) *School leadership and system reform in the 21st Century*. London: Bloomsbury.

Eco, U. (1979) *A theory of semiotics*. Bloomington, IN: Indiana University Press.

Fullan, M. (2011) *Change leader: Learning to do what matters most*. San Francisco: Jossey-Bass/Wiley.

Gazendam, H., Jorna, R. and Cijsouw, R. (2003) *Dynamics and change in organizations: Studies in organizational semiotics*. New York: Springer.

Greany, T. and Higham, R. (2018) *Hierarchy, markets, networks, and leadership: understanding the 'self-improving school system' agenda in England*. London: UCL Institute of Education Press.

Hairon, S. and Goh, J. (2015) Pursuing the elusive construct of distributed leadership: is this search over?, *Educational Management & Leadership*, 43(5): 693–718. https://doi.org/10.1177/1741143214535745

Hauge, T.E., Norenes, S.O. and Vedöy, G. (2014) School leadership and educational change: Tools and practices in shared school leadership development, *Journal of Educational Change*, 15: 357–76. https://doi.org/10.1007/s10833-014-9228-y

Kahneman, D. (2011) *Thinking, fast and slow*. New York: Farrar, Straus and Giroux.

Kotter, J. (2014) *Accelerate: Building strategic agility for a faster-moving world*. Boston: Harvard Business Review Press.

Lai, E. and Cheung, D. (2014) Enacting teacher leadership: The role of teachers in bringing about change, *Educational Management Administration & Leadership*, 43(5): 673–92. https://doi.org/10.1177/1741143214535742

Lazarsfeld, P.F. and Merton, R.K. (1954) Friendship as a social process: a substantive and methodological analysis, *Freedom and Control in Modern Society*: 18 (1): 66.

Lee, M. and Louis, K.S. (2019) Mapping a strong school culture and linking it to sustainable school improvement, *Teaching and Teacher Education*, 81: 84–96. https://doi.org/10.1016/j.tate.2019.02.001

Leithwood, K. and Louis, K.S. (2012) *Linking leadership to student learning*. San Francisco: Jossey-Bass.

Lukacs, K. (2012) Exploring "The Ripple in the Pond"- A correlational study of the relationships between demographic variables and the teacher change agent scale, *Current Issues in Education*, 15(2): 1–11. https://doi.org/10.1177/1541344616655887

Marzano, J., Waters, T. and McNulty, B. (2005) *School leadership that works: From research to results*. Alexandria, VA: ASCD.

OECD (2016) *What makes a learning organization: a guide for policy makers, school leaders and teachers*. Available at: http://www.oecd.org/education/school/school-learning-organisation.pdf (accessed 24 June 2023).

Poekert, P., Alexandrou, A. and Shannon, D. (2016) How teachers become leaders: an internationally validated theoretical model of teacher leadership development, *Research in Post-Compulsory Education*, 21(4): 307–29.

Puccia, E., Martin, J.P., Smith, C.A.S. et al. (2021) The influence of expressive and instrumental social capital from parents on women and underrepresented minority students' declaration and persistence in engineering majors, *International Journal of STEM Education*, 8: 20. https://doi.org/10.1186/s40594-021-00277-0

Robinson, V., Hohepa, M. and Lloyd, D. (2008) *School leadership and student outcomes: Identifying what works and why: Best Evidence Synthesis iteration [BES]*. Wellington, NZ: Ministry of Education.

Robinson, V.M.J., Lloyd, C.A. and Rowe, K.J. (2008) The impact of leadership on student outcomes: an analysis of the differential effects of leadership types, *Educational Administration Quarterly*, 44(5): 635–74. https://doi.org/10.1177/0013161X08321509

Rogers, E. (1995) *Diffusion of innovations*, 4th edn. New York: The Free Press.

Rose, J., Thomas, S., Zhang, L. et al. (2017) Research learning communities: Evaluation report and executive summary. Available at https://educationendowmentfoundation. org.uk/public/files/Projects/Evaluation_Reports/Research_Learning_Communities. pdf (accessed 2 November 2022).

Schildkamp, K. and Datnow, A. (2020) When data teams struggle: Learning from less successful data use efforts, leadership and policy in schools. https://doi.org/10.1080/1 5700763.2020.1734630.

Schnellert, L. (2020) *Professional Learning Networks: Facilitating transformation in diverse contexts with equity-seeking communities*. Bingley, England: Emerald.

Tulowitzki, P. and Pietsch, M. (2018) *The differential and shared effects of leadership for learning on teachers' organizational commitment and job satisfaction: A multilevel analysis*. Presented at the European Conference on Educational Research annual meeting, Bolzano, Italy, 4–7 September.

Tversky, A. and Kahneman, D. (1974) Judgment under uncertainty: Heuristics and biases, *Science*, 185(4157): 1124–31. https://www.science.org/doi/10.1126/science.185.4157.1124

Von Hippel, A. (2014) Program planning caught between heterogeneous expectations – An approach to the differentiation of contradictory constellations and professional antinomies, *Edukacja Dorosłych 2014*, 1(70): 169–84.

Warren Little, J. (1990) The persistence of privacy: Autonomy and initiative in teachers' professional relations, *Teachers College Record*, 91(4): 509–35. https://doi.org/10.1177/ 016146819009100403

Watson, C. (2014) Effective professional learning communities? The possibilities for teachers as agents of change in schools, *British Educational Research Journal*, 40(1): 18–29. https://doi.org/10.1002/berj.3025

Wenner, J.A. and Campbell, T. (2017) The theoretical and empirical basis of teacher leadership: A review of the literature, *Review of Educational Research*, 87(1): 134–71. https://doi.org/10.3102/0034654316653478

Index